Beginner's Guide to Programming the PIC32

By Thomas Kibalo

Published by Electronic Products.
Copyright 2013, Electronic Products (USA). All rights reserved.

No part of this publication may be reproduced, stored in a retrieval system, or transmitted in any form or by any means, electronic, mechanical, photocopying, recording, or otherwise, without the prior written permission of the publisher.

The publisher offers special discounts on bulk orders of this book.

For information contact:

Electronic Products
P.O. Box 251
Milford, MI 48381
www.elproducts.com
chuck@elproducts.com

The Microchip name and logo, MPLAB® and PIC® are registered trademarks of Microchip Technology Inc. in the U.S.A. and other countries. PICkit™ is a trademark of Microchip Technology Inc. in the U.S.A. and other countries.

All other trademarks mentioned herein are property of their respective companies.

Printed in United States of America
Cover design by Rich Scherlitz

This book is dedicated to the memory of my parents John and Betty Rose Mote Kibalo. Thank you for all the love and kindness you shared with me, it continues to fortify me and guide me throughout my life. Rest well, dear ones, in a job well done.

"What drives and validates self-help is a belief, that our energies and abilities are finally congruent with the universe, that our minds, are at the bottom a match for the nature of things"
William James

Table of Contents

Introduction 9

Chapter 1 – Getting Started 11
    Introducing Microchip MPLAB X IDE 11
    Software Installation 13
    Steps to install MPLAB and Compiler 14
    Build the Demo 24
    Assembling the Microstick II to a Solderless Breadboard 25
    Connecting the Microstick II to your PC 27
    Running the Demo 28
    Microchip 32 bit Microcontroller Family 31
    The PIC32MX250F128B Microcontroller 33
    Exercises: 38

Chapter 2 – Driving an LED Display 41
    PIC32MX Programmable Port overview 41
    PIC32MX IO using Peripheral Library 43
    Experiment 1- Hardware 45
    Experiment 1- The Software 46
    A Library versus Non-Library Implementation 55
    Experiment 1- How it Works 56
    Experiment 1 Execution 56
    Experiment 2 65
    Review of PIC32MX250F128B Key features: 69

Chapter 3 – Reading an Input Switch 73
    Experiment 1- Hardware 74
    Experiment 1- The Software 78
    Experiment 1 Execution 79
    Experiment 2 -The Hardware 82
    Experiment 2 –How it works 83

Experiment 2 -The Software 83
Experiment 2 Execution 86

Chapter 4 - PIC32 Interrupts and Change Detection 89
The PIC32MX250F128B Interrupt Capability 90
The PIC32 interrupt ISR 92
Designing for interrupt services 93
Exercise 1 INT0 Interrupt 93
Exercise 2 Change Notice (CN) Interrupt 96

Chapter 5 – Using the ADC 103
Hardware Description 109
Software Description 110
Running the Experiment 113

Chapter 6 – Using an LCD Display with the Microstick II ... 119
Hardware Description 119
Software Library Description 122
Experiment #1 – LCDDEMO 126
Running LCDDEMO 127
Experiment #2 –A Simple Voltmeter 128
Running the Voltmeter 130

Chapter 7 – Using Timers and Timer Interrupts 133
The 16 bit Timer Peripheral 134
Hardware Description 135
Timer 1 Experiment –Non-Interrupt 137
Timer 2 - 16 bit Execution 140
Timer 2 - Experiment using Interrupt 140
The 32 bit Timer Using Timer Peripheral Pairs 143
Timer2/3 32 Bit Operation –Non Interrupt 144
Timer2/3 32 Bit Operation Interrupt 145

Chapter 8 - Optimizing PIC32 Performance 149
MIPS 32 4K Core Instruction and Data Operations 149
MIPS 32 4K Internal Operations 151
First Optimization- Cache and Wait State Configuration 152
Second Optimization –System Clock 153

    Third Optimization – Compiler Settings 156
    Exercises.. 156

Chapter 9 – Serial RS232 Communications............................ 163

    Serial Communications Overview 163
    Industry Standard External Wired Serial Communications 166
    USB (Universal serial Bus) to PC Connectivity 169
    American Standard Code for Information Interchange (ASCII).................................. 170
    The PIC32MX Peripheral Programming System (PPS) 172
    Initializing PPS.. 174
    A UART Application Software Library... 179
    Experiment 1 "Implementing a RS-232 Serial interface with a PC 180
    Hardware Configuration 180
    Bringing up HyperTerminal 182
    Executing the code 184
    Using USB instead of RS-232 186
    Bringing up Tera Term 188
    Experiment 2 "Using Interrupt capability with UART2"... 190

Chapter 10 – The Synchronous Peripheral Interface (SPI).. 195

    Introducing a Serial EEPROM... 197
    Experiment 1- Hardware... 198
    Experiment 1- The Software an EEPROM Library 200
    Experiment 1- How it Works 202
    Experiment 1 Execution 205

Chapter 11 – Using PWM for Tone Generation..................... 209

    Output Compare Module (OCx)... 210
    Exercise 1 LED PWM Experiment... 211
    Software operation.. 214
    Running the Light Dimmer Experiment... 216
    Exercise 2 Tone Generation 217
    Software Overview... 217
    Running the Sound Experiment 219

Chapter 12 –RTCC (Real Time Clock Calendar) 223

    The Real Time Clock Calendar 223

Configuring and using the RTCC 226
Experiment 1 -Date and Time Setting and Operation 228
Experiment 2 Date and Time with LCD 232
Experiment 3 Alarm and Interrupt 235

Chapter 13 –Executing Arduino Code on PIC32 241

Core Subset of Arduino Library Functions 242
The PIC32MX250F128B Hardware Model 243
Arduino Blink Example #1 245
Running Example 1 Code 247
Arduino Digital Serial Read Example #2 249
Running Exercise 2 Code 252
Arduino Analog Serial Read Example#3 255
Running Example 3 Code 257
Arduino Fading Example #4 259
Running Example 4 code 261

INDEX 267

Introduction

Microcontrollers have made their mark on the modern world, and in fact, are helping to redefine it. New microcontroller apps are emerging, making increasing demands on the microcontroller for low power, greater real time performance, and increased memory size, and low cost. The good news is that with 32 bit microcontrollers like the Microchip PIC32, microcontrollers can now handle these challenges as well as execute multiple applications simultaneously. Using a PIC32 microcontroller, an application can simultaneously incorporate high resolution color graphics with touch screen, mobile Wi-Fi internet based connectivity and Host USB 2.0 for reading and writing to Flash drive USB Stick. The entire application can run with a real time operating system (RTOS) while still maintaining battery operations. The application possibilities for the PIC32 seem truly endless.

In this beginner book we will explore the basic concepts to allow you to gain familiarization with the PIC32MX processor and its tools suite. Our approach will first be to install and become familiar with the Microchip tools suite using the Microstick II, with PIC32MX, as a programmer debugger. We will then embark on the use of digital I/O for LED activation and input switch detection, conduct important "hands on" exercises for ADC, SPI, PWM and RTCC, to increase our knowledge of the tools suite and the associated PIC32MX hardware. We will cover in some detail the internal PIC32MX CPU and its interrupt architecture and capabilities and learn techniques to maximize its performance. As the book concentrates on a "hands on" approach (where each chapter is experiment based and a prototype is developed) some basic experience with electronic prototyping as well as some familiarity with solderless bread boarding techniques is assumed. Each chapter introduces a new component with the microcontroller and leverages on the knowledge and experience gained from the previous chapters. You are encouraged to work the book from beginning to end. All chapters end with a review and exercise section to help reinforce and challenge you on the knowledge gained in that chapter.

All these designs are done in 'C' code using the Microchip free software XC complier, and so some familiarity with 'C' is assumed. The 'C' language has emerged over the use of assembly language as the predominant standard for the new generation of 32 bit machines. This is because of the inherent portability of 'C' code across all microcontroller types and families, the availability of a large open source of 'C' code and free application libraries, and the fact that these new machines are powerful enough to handle real time operation with 'C' without performance impact.

To speed up the learning cycle you will not be required to write any code on your own, all experiments will have proven code in place. In addition all designs will also be thoroughly discussed, and the development tools will be covered in sufficient detail to allow you to embark on your own. In the end you should be enabled to modify the code provided, and use these exercises to develop your own original code.

My sincere hope is that upon reading and utilizing the material in this book you will gain the necessary confidence with these remarkable devices to move forward with their use in your own applications. Microcontrollers are an exciting and challenging area to work, requiring a mix of both hardware and software skills. The good news here is that the development tool sets to deal with this challenge are also improving. The Microchip Integrated development environment (IDE), MPLAB X, and the Microstick II programmer/debugger are examples of just such tools.

The new Microchip IDE MPLAB X is an award winning environment that links all the critical development tools for the embedded design cycle within a single GUI framework. MPLAB® X Integrated Development Environment brings many changes to the PIC® microcontroller development tool chain. Unlike previous versions of MPLAB® which were developed completely in-house, MPLAB® X is based on the open source Net Beans IDE from Oracle. Taking this path has allowed Microchip to add many frequently requested features very quickly and easily while also providing a much more extensible architecture to bring users even more new features in the future.

MPLAB-X supports hardware tools like the Microchip's Microstick II that allow "In Circuit Programming" and "Real Time Debug" to enable both programing and debugging of the microcontroller while it is still in prototype stage.
In this beginner's book we guide you through embedded design process using the Microchip IDE MPLAB X and the Microstick II as our programming/debugging tool. We will focus on using this tool suite with a PC running Windows operating system. We will also utilize the PIC32 Microchip 32 bit Microcontroller PICMX2 series for all of our experiments.

The Arduino has emerged as a microcontroller hardware and software standard, among the hobbyist community. In recognition of this, and to help bridge the gap between this standard and the PIC32 hardware and software environment, a chapter in this book is devoted to using an Arduino Standard Reference library project under MPLAB X for the PIC32MX. The project will serve as a core template to repeat several of the PIC32MX exercises covered in this book. By doing so, an experienced Arduino user can also gain immediate appreciation for the use of PIC32MX and what it has to offer for an existing Arduino application, along with the use of the enhanced Microchip Development MPLAB X environment.

Chapter 1 – Getting Started

In this chapter we will guide you through the installation process for Microchip MPLAB X, the Microchip XC32 'C' compiler, and the Microstick II programmer/debugger. We will test this installation using the demo project "BlinkLED" to help insure that all elements of the installation are correct and complete. We will then cover some of the important hardware elements of our tool suite, starting with an overview of Microchip's 32 bit Microcontroller family, and then covering in more detail the PIC32MX250F128B (the PIC32MX part used with the Microstick II), and then finally discussing the Microstick II and its operation.

This chapter will serve as the foundation for all other chapters, and will leave us with the necessary development tool suite, and hardware overviews, for all subsequent chapter discussions and experiments.

Introducing Microchip MPLAB X IDE

The Microchip MPLAB X IDE is multi-platform that runs under Windows, Mac, or Linux operating systems and is used to develop applications for Microchip microcontrollers. As an IDE, it provides a single integrated "environment" to develop and test code, as part of the embedded design cycle required for all microcontrollers. The environment assists the designer in progressing through the "embedded design cycle"(as shown in Figure 1-1) without the distraction of switching among an array of distinct and different tools, and allows for the designer to concentrate on completing the application without the interruption of using these separate tools and their associated specific different modes of operation.

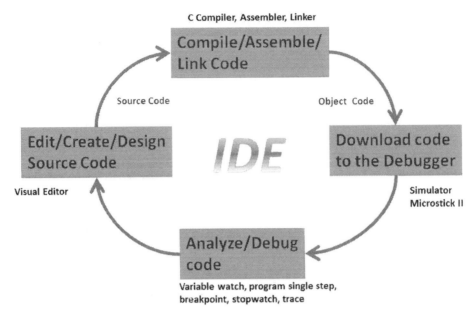

Figure 1-1: Embedded Design Cycle

The MPLAB X IDE, as a framework, coordinates all the tools from a single graphical user interface. For instance, once source code (i.e. 'C' code) is written, it must be converted to executable instruction set through conversion to assembly code and linked to other assembly code or libraries (object code) and then downloaded and programmed into a microcontroller to determine how well it works. During this process multiple tools are needed: an editor to write the code, a project manager to organize files and settings, a compiler to convert the source code to object code, a linker to combine this object with existing libraries, and then a hardware programmer/debugger to test the object code in operation with the microcontroller. MPLAB X along with the Microstick II programmer/debugger meets all these requirements, and will serve in this book as our basic development environment.

The basic components and features of MPLAB X are shown in the Figure 1-2. In the chapters to follow we will continue to discuss and illustrate the detailed use of this integrated tool set as a critical part of each chapters' "hands on" experiments.

- A mature feature-rich IDE IDE
- Cross platform compatibility (LINUX, WINDOWS)
- Open source
- Support for advanced high level language development
- A large ecosystem of plug-ins
- An IDE that keeps track of changes
- A powerful project navigation system

Figure 1-2: MPLAB X Features and Functions

Software Installation

The software installation of MPLAB couldn't be easier. All software and tools are available for download from Microchip's web site (the link is provided). The demo code for this book, the MPLAB project content, and as well as an experimenter kit for all electronics parts used throughout the book exercises (exclusive of the Microstick II) is also available from the author's web site (www.Kibacorp.com). Our first goal is to create for ourselves an integrated development environment that looks like Figure 1-3. Let's get started.

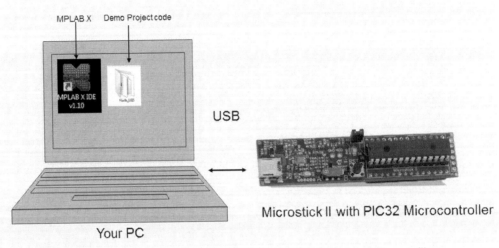

Figure 1-3: Development Environment

Steps to install MPLAB and Compiler

Before you can use MPLAB X IDE you must do the following:
- Install JRE (Java Runtime Environment) and MPLAB X IDE
- Install USB Device Drivers (For Microstick II)
- Connect to a Target (Using Microstick II)
- Install the Language Tools
- Install chapter 1 demo code
- Launch the IDE

We will cover each step here.

1. **Install MPLAB IDE and JRE (Java Runtime Environment) on PC**
   - You can find MPLAB X here:
     http://www.microchip.com/mplabx/
   - Click MPLAB X Free Download button.
   - The Microstick II, our programmer/debugger, is designed run with MPLAB X version 1.00 or later. Make sure the MPLAB X is version is at least 1.00
   - The install is very straight forward. Follow the prompts, restart your PC then proceed.
   - MPLAB X does require your PC to have Java Runtime Environment (JRE). Ensure you have the correct JRE for your version of MPLAB X

IDE by consulting the "Readme for MPLAB X IDE" accessed from the Release Notes on the Start page. Read the following sections and print them out if you need to perform the actions specified in them. Close MPLAB X IDE until the required steps are performed and then re-launch
- You should end up with MPLAB installed on your hard drive under C:/Programs/Microchip/MPLAB directories
- You should also end up with an MPLAB short cut icon on your desktop.

2. **Install the USB device drivers (For Microstick II)**
   For correct tool operation Microstick USB drivers will need to be installed. This happens automatically with the Microstick II once the Microstick II is plugged into the USB port of your PC. The Microstick II is factory populated with the PIC32MX 32 bit microcontroller. The specific microcontroller we use is the PIC32MX250F128B

3. **Install the Microchip XC32-Compiler on PC**
   - You can find it here or access it directly from the start page by selecting "download compilers and assemblers".

   - http://www.microchip.com/mplabxc/

   - The install is very straight forward. Follow the prompts. Click downloads and then select XC32 for Windows
   - Download the compiler code into any convenient directory on your PC and double click to install.
   - After installation you should end up with the XC32 Complier installed on your hard drive under C:\Program Files (x86)\Microchip\xc32\v1.00(or later)
   - If you navigate to that directory, you should see the folder set shown in Figure 1-4.

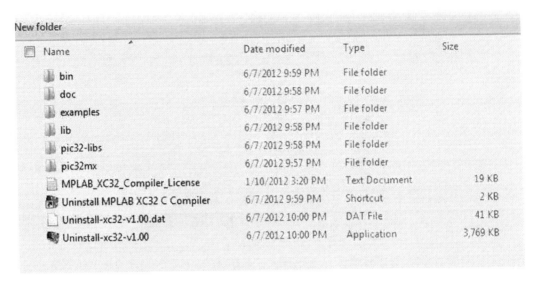

Figure 1-4: MPLAB C32 Directory

FYI
PIC32 parts support and a list of complier documents should have been automatically installed under the doc folder including user guides, tools libraries and library help. Other information is available as part of the MPLAB X readme document located on Microchip Web site.

- *MPLAB® XC32 Compiler for PIC32 MCUs User's Guide (DS51686E - updated for v1.10)*
- *MPLAB® 32-Bit Language Tools Libraries (DS51685)*
- *Microchip PIC32MX Peripheral Library - Compiled Help (CHM)*
- *MPLAB® Assembler, Linker and Utilities for PIC32 MCUs Users Guide (DS51833)*

4. **Install the Demo Code**

   You can find the demo code for all chapters in the book at the following URL http://kibacorp.com/free-downloads.
   Install it either on your desktop or a convenient place on your hard drive

5. **Launch the IDE**

   Let's give these entire install a "test drive". A good first place to start is just to 'fire' up MPLAB X itself and see what happens. Double Click the MPLAB X icon on your desktop.

The MPLAB IDE start page should come up as follows:

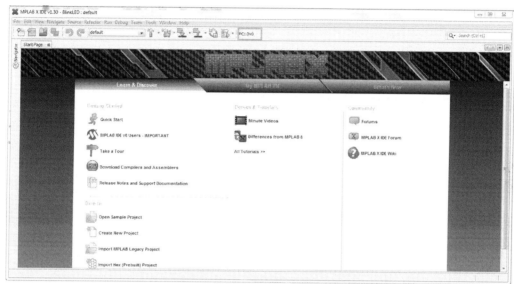

Figure 1-5: MPLAB X Start Page

On the start page there are three tabs
- Learn and Discover
- My MPLAB IDE
- What's New

Each of these are very much self-explanatory, and details can be found by referencing the MPLAB X users Guide available under the Getting Started Release Notes and Support Documentation or just simply experimenting on your own, but let's move on.

Ok so far, but not very exciting. The MPLAB works well –but where is the beef? Let's open a Demo project directly under MPLAB X. We should see MPLAB populated with the components of the particular demo project, displaying those functional components in MPLAB in the workspace as they were last saved.

Leave MPLAB X Open, and click "File" on tool bar, from the pull down menu "select open project". An Open Project Dialog Box appears. Navigate to where you stored your demo code. Open Chapter One folder, then locate folder

BlinkLED. In this folder you should see the BlinkLED.X, an MPLAB X project icon. Click on this icon. Figure 1-6 captures the whole process.

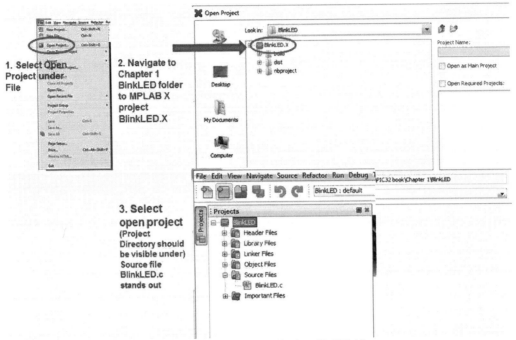

Figure 1-6: Using Demo Code with MPLAB

Once this icon is selected the IDE view will change. The IDE will show a project view is on the left hand side of the IDE. It contains all the code files and files directories possible within an MPLAB project. The directories may or may not be populated. It is up to you the designer as to what code you put into your project. For this demo all we have is single 'C' source code file. Open the project view source code directory by clicking on the '+' symbol next to Source files. Its contents should be a BlinkLED.c file. Double click on this source code icon in project view. You should see what is shown in Figure 1-7.

Figure 1-7: Viewing Source Code

The start page is now a tab and a new page is now added showing the project source code now opened under an Edit Window. The source code is color coded according to 'C' code syntax type and is completely viewable using the scroll bar. Now let's add to the IDE more window content for a complete picture. Select navigator and a window should automatically open in lower left hand corner of the IDE. This window shows all the critical code elements in the present source code view. In this case we should see the image in Figure 1-8.

Figure 1-8: Navigator Panes

Beginner's Guide to Programming the PIC32

Clicking on any icon reference in this window will cause the source code editor cursor to line up directly with the icon reference. The revised IDE should look as shown in Figure 1-9 with Navigator in lower left hand corner.

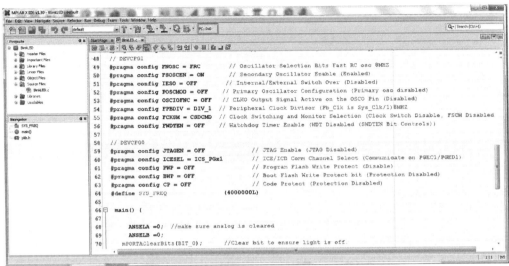

Figure 1-9: IDE with Project, Navigator, and Source views.

The next window we will open is the dashboard. Select Window from the Main Tool bar and from pull down menu, Dashboard. You should see the following:

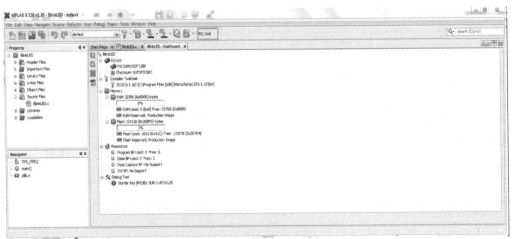

Figure 1-10: Dashboard

The dashboard is an on-demand complete synopsis of the project. It also allows for quick view of project properties –all very cool:

Beginner's Guide to Programming the PIC32   Page 20

- Device –The specific Microchip microcontroller the project was built for, in our case it is a PIC32, the PIC32MX250F128B.
- Compiler Tool chain- The specific tool compiler that the code was compiled with, in our case Microchip XC32.
- Memory- This is a tally of the PIC32MX250F128 internal FLASH and RAM used by the project code
- Resources-The number of breakpoints used by the project and what is totally available, more on this later.
- Debug Tool-This is the specific hardware debug tool used by the project. In our case we have a Microstick II Starter Kit, and the serial number associated with last used Microstick II.

There are a number of right side icons associated with it to assist with Dashboard functionality.

Icon	Function
	Display the Project Properties dialog.
	Refresh debug tool status. Click this to see hardware debug tool details.
	Toggle software breakpoint. Click to alternately enable or disable software breakpoints.
	Get device data sheet from Microchip web site. Click to either open a saved, local data sheet or open a browser to go to the Microchip web site to search for a data sheet.

Figure 1-11: Dashboard Icons

To change Project Settings (i.e. Processor, Compiler, debug tool) just click the "wrench" icon and you will get the Project Properties Dialog Box. As shown below. Here configuration changes can be done. None are currently required as the project has already been pre-built for you. Simply hit cancel and return to Dashboard. To refresh the Dashboard contents at any time hit the "refresh icon".

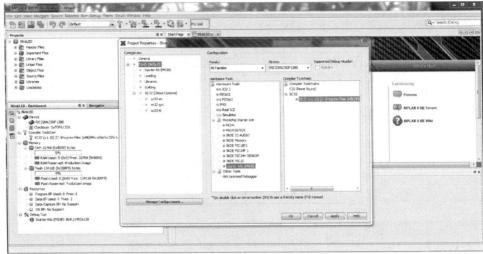

Figure 1-12: Changing project properties via Dashboard

A very convenient feature with the Dashboard is the on-demand view of the project's target microcontroller datasheet, in our case; this is the Microstick II PIC32MX250F128B. The first time you use this you will have to navigate for the data sheet as to where it is on your PC. After that the IDE project will retain the location.

Notice that the dashboard window it is tabbed with the other windows (source code editor and start page). Only one window can be viewed at a time. You can alternate your views by clicking the associated tab.

For convience the Dashboard tab can also be clicked and dragged into the Navigator view to create a multi-tab view in that area.

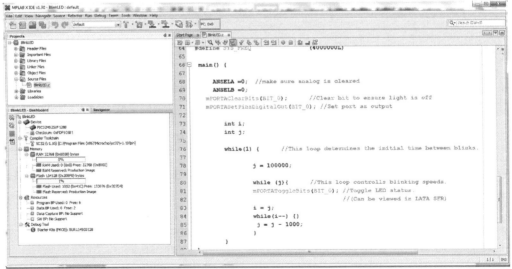

Figure 1-13: Incorporating Dashboard into Navigator Pane

Beginner's Guide to Programming the PIC32

Before moving on let's close with a final top level view of the IDE with different tools bars and window panes that we can refer to through out the book. The MainTool bar and Editor tool bars are shown in Figure 1-14.

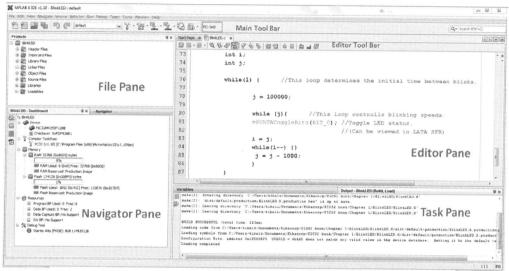

Figure 1-14: MPLAB X Top level view of Panes and Tool bars

- File pane – A pane with up to four file-related tabbed windows. The Projects window displays the project tree, the Files window displays the project files, a Classes window can display any classes in the code, and a Services window displays any services available to use for code development.
- Navigator pane – A pane that displays information on the symbols and variables in the file selected in the File pane.
- Editor pane – A pane for viewing and editing project files. The Start Page also is visible here.
- Task pane – A pane that displays task output from building, debugging or running an application.

If you double click on any file name in the File pane, the related file will open in the Editor pane under a tab next to the Start Page. To close the tab, click on the "x" next to the file name. Right click on the project name in the File pane, Projects window, to view the pop-up (context) menu. Do the same for the project's subfolders.

As you now realize there is a lot of capabilty built into the MPLAB X, and throughout this book we will be workly closely and hands on with most of its content. Since all of the demostrating excercises in this book are already pre-built projects we thought it appropiate to include a built from scratch project excercise using MPLAB X in Appendix A to give you that hands-on as well. More detailed information on MPLAB X is available as document 52027A the MPLAB X Users Guide.

Build the Demo

Let's now move on. Unlike MPLAB 8.XX with the MPLAB X IDE it is not necessary to build the project first and then run or debug. Building is an automatic part of the MPLAB X run and debug processes. For our initial development, however, we want to make sure that the project builds before attempting to run or debug.

There are several options available to build a project.
- In the Project window, right click on the project name and select "Build". You may also select "Clean and Build" to remove intermediary files before building.
- Click on the "Build Project" or "Clean and Build Project" toolbar icon.

 Build Icon

 Clean and Build Icon

Click now on the Build and Make Tool Bar or "hammer icon" in the tool bar. This should populate the output window with the results of the build (successful build). Build progress will be visible in the Task Pane Output window (lower right hand corner of the desktop). For our demo, the code should build successfully.

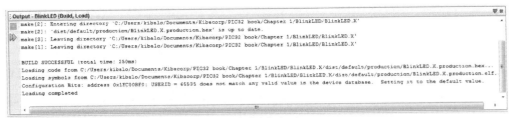

Figure 1-15: Build Output

Now we have MPLAB X open with a project window, source code window, Dashboard/navigator window and output window. The Output window should indicate a successful build. The IDE should now appear as shown in Figure 1-16.

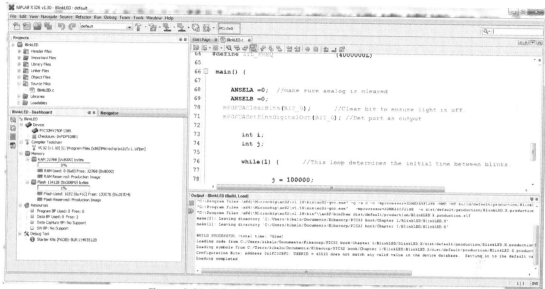

Figure 1-16: IDE with Output shown

Assembling the Microstick II to a Solderless Breadboard

Before executing the demo we need to assemble and then connect our Microstick II to the PC. This section just focuses on the assembly and mounting the assembly onto a Solderless breadboard. Pictures are worth "a thousand words", Figure 1-17 contains a picture of the Microstick II "Out of the Box "with both top and bottom views show the .100 headers alongside both views of the board.

Figure 1-18 shows the Microstick II side view highlighting how the .100 " headers are inserted into the bottom socket and then how this entire assembly is plugged into a Solderless breadboard. The one shown here is a medium size that can facilitate smaller prototypes. For those unfamiliar with this Solderless technology it allows for quick point to point wiring assembly of electronic prototypes using through hole technologies (i.e. Dual In Line Package (DIP) circuits, resistors, LEDs, switches, displays, capacitors). The Microstick II essentially functions as the DIP microcontroller in this setup.

When using Solderless breadboards you need to place the DIP technology over the column gap. This DIP placement allow for pins on either side of the DIP to not be shorted together and also have access to a full 5 pin column of connection that is unique for each pin. These column connections allow for easy connection to other electronics with the DIP. In addition there is a common bus that is configured as a row structure allowing easy distribution for power and ground to other electronics on the breadboard.

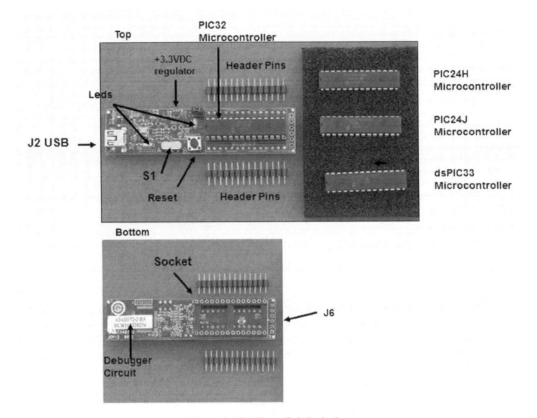

Figure 1-17: Microstick II photos

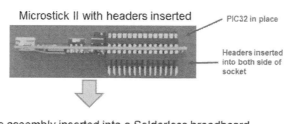

Figure 1-18: Microstick II plugging into Solderless breadboard

Note that with the Microstick II as installed in these photos and given its board size of 20x76 mm, column pin access is limited to two connections each. This is more than enough for most prototyping efforts.

Connecting the Microstick II to your PC

Please connect your Microstick II at this time, if it hasn't been connected already. A Microchip USB driver is automatically installed the first time you hook up your Microstick II. Once done the Windows operating system should recognize the Microstick II connection automatically afterwards. Once the Microstick II is connected and driver installed the Microstick II Debug LED green should be constantly on. To confirm MPLAB connection click refresh on Dashboard and the output window task pane should show the following. The Dashboard should also update with voltage level to the Microstick as well as Microstick type, serial number, device type and current revision numbers.

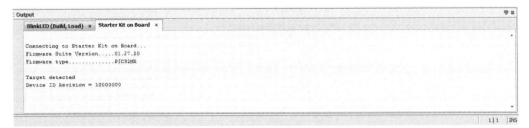

Figure 1-19: Verifying Microstick connectivity

Running the Demo

The Microstick II should now be on a Solderless breadboard and powered via USB connection to your PC with the Microstick debug LED is on steady.
At this point it may appear that you are almost done, but we need to take it to a final step. For this final step we just need to get the IDE development system in concert with the prototype to actually do something useful. The project, when it finally runs on the Microstick II, blinks the only available user LED that is located on the board. Let's prove it.

First take a closer look at the Build tool bar. First take a closer look at the Build tool bar (Figure 3-20). It will be an instrumental tool in running the demo; you need to become familiar with it as we will be referring to this tool bar and its components when executing the demo.

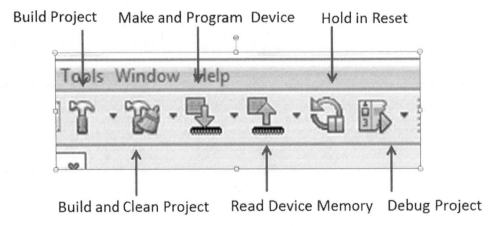

Figure 1-20: Build Tools

For this demo, the code used has been tested and runs. However, as you develop your own code it may need to be debugged as you develop your application, so let's run the debugger.

To run the demo code under debug, click on the "Debug Project" icon.

 Debug Run Icon

It will cause a build if the make process determines it is necessary and then program and run the PIC32 on the Microstick.

Position your mouse over the tool bar to "Select Build" button or "Hammer icon You should see a successful build (which follows a stream of text build activities that occur in output window) and execution; this is the absolute final step in the process. Here's where the payoff for all our hard work occurs. You should see a blinking LED.

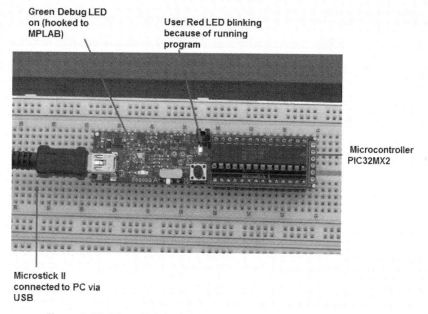

Figure 1-21: Microstick Prototype running BlinkLED demo

The Microstick II should be running and the IDE should be responding with a visible activity bar active in the lower left hand corner of the IDE window. In addition, the run button on tool bar is disabled, and only the pause button is available for use. There is a new active tool bar once Debug Project is selected

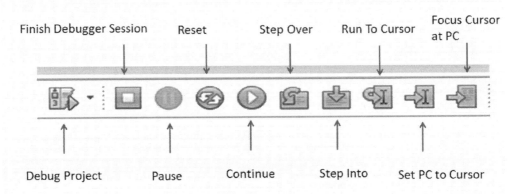

Figure 1-22: Debugger tool bar

You can click pause button on the tool bar and the Microstick II will stop at the current line that the microcontroller was beginning to execute. A green arrow will appear in the source code window (if it is open) as to what line of source the microcontroller is paused at. You can then elect to run from there (click continue), or use single step (step over) to execute one line of source at a time, or reset (click reset) and the Microchip will start execution of your code from the beginning.

It would help here to briefly cover some of the important debug functions, discussed so far, that reside in the IDE tool bar (Figure 1-20 and Figure 1-22)

- Build Project – just rebuild code that is new
- Make and Build All – rebuild all code in Project Window
- Make and Program- the device (non-debug mode target can run on power up without IDE. This is a release mode.
- Read the Device- reads device memory for viewing under IDE
- Reset –hold the device in reset. Once released execution will occur from beginning of code
- Debug Run –execute the program under debug
    - Pause- stop the program
    - Continue- resume program execution
    - Reset- starts debugger at beginning of program
    - Single Step (step over, step into) – execute one line of code from where you previously paused and then pause after execution. Step

over allows current function or line to be executed. Step into, if at a function, will travel into and out of that function during single step.
- Run to cursor – run program and then break at cursor location in editor pane.
- Finish Debug- exits debug mode, turns off Microstick power.

Congratulations! –you now have a fully functional microcontroller development system! With this tool suite you can now undertake the exercises in this book and begin to progressively prepare and train yourself in 32 bit Microcontroller embedded design techniques.

We will be covering the function and features of MPLAB X more carefully in the chapters to follow. Now let's discuss some of the important hardware features of the Microstick II and the PIC32MX2 microcontroller so as to better understand these important resources and their full capabilities.

Install Errors

1. <u>'C' Complier Not installed</u> –MPLAB-X is aware of all tools installed within your machine. It can co-exist side by side with older versions of MPLAB 8.xx series without issue. You can always check MPLAB-X available tools by looking at the dashboard display. A "green" indication next to the tool indicates MPLAB-X can access it, our project was built using XC32 compiler. If not available then you need to install following install instruction given earlier in this chapter. IDE will remember where it is.
2. <u>Debugger not connected</u>-If IDE cannot find Microstick II debugger then it will prevent a debug or program actions to occur outside of just using MPLAB simulator. Please connect the Microstick II to your PC at an available USB port. The IDE as it reports through the Output Window should recognize the connection.

Microchip 32 bit Microcontroller Family

Microchip offers a number of 32 bit solutions for its microcontrollers (see Figure 1-23). The PIC32MX2 microcontroller represents the entry level of the PIC32 family and it incorporates the maximum number of peripherals on chip, but at the same time, the lowest MIPS rate.

All of the PIC32 series are +3.3V logic. For this book we will focus on experiments and operation for only the PIC32MX2.

The Microstick II is socketed and can support the PIC32 (32 bit microcontrollers), PIC24H, PIC24J and dsPIC33 (16 bit microcontrollers) targets. It ships with the PIC32 in the socket. The socket only supports 28 pin skinny DIP packages. Your Microstick II ships with four such devices, the PIC32MX250F128B, the PIC24HJ128GP502, the dsPIC33FJ128MC802, and the PIC24FJ64GB002. All of these parts are +3.3VDC logic and have a similar pin out configurations. The PIC24H and dsPIC33 have memory capacities of 128K Flash and 8K Ram and both can operate up to full rate of 40 Million Instructions per Second (MIPS). The PIC24F has a reduced memory capacity of 64K Flash and 8K Ram with a 32 MIPS rate.

Why use a Microchip 32 bit part? After all, the large variety of Microchip 8 bit parts have been around for a while and are all well supported. To facilitate this decision it may help to carefully review both performance and capability of a PIC32 32 bit processor against 16 bit processor side by side with an 8 bit machine. A table is provided to illustrate this point. Let's compare the PIC32M2 part to PIC24F to a well-known 8 bit member of the Microchip family the PIC16F887.

PIC32MX2	PIC24F	PIC16F
32 bit	16 bit	8 bit
128K Flash	64 K Flash	8k Flash
32K RAM	8K RAM	386 RAM
40 MIPS	16 MIPS	5 MIPS
3.3V operation	3.3v operation	5 V operation

We will compare data size, memory size, clock rates and final instruction speed measured in MIPS (Millions of Instructions per Second). The comparison shows that the PIC32MX has 16x the memory capacity in flash, over 80x the RAM capacity, and a MIPS rate of 8x the PIC16F887. If we also compare peripheral content we will find that the 32 bit exceeds by a factor or 2:1 numerically over most of the 8 bit internal peripherals, and introduces new peripherals. Finally as the 32 bit data bus is 4x the size of the 8 data bus meaning that four as much data is handled during any one instruction. With this type of processing power and peripheral power you can handle a broader range of applications that 8 bit world could not handle standalone. We discussed these briefly in the introduction but could include network connectivity with TCP/IP stack, high resolution graphics, wireless stacks like ZIGBEE, 802.11b, SD-CARD with 2GB+ capacity that is formatted in a Windows compatible file I/O, to name a few.

The PIC32MX250F128B Microcontroller

Let's dig a little deeper and now explore the microcontroller PIC32MX250F128B more closely. This device, as mentioned earlier, comes prepackaged in a "skinny DIP". As discussed earlier the PIC32MX250F128B is a +3.3VDC logic part and supports 40 MIPS of operation with both a 32K FLASH, 8K RAM.

Interrupt handling and response times are important considerations in the use of microcontrollers for any real time operations. The PIC32MX2 is well equipped here. It has an internal interrupt controller with guaranteed 5-cycle latency in response to interrupts, up to 45 available interrupt sources, with seven priority levels, five traps for any processor exception handling (i.e. divide by zero) and three external interrupts.

A high level block diagram of the PIC32MX2 microcontroller is shown in Figure 1-23. The chip is organized around two internal buses. The top bus is the faster of the two and runs at the system clock CPU rate. It allows simultaneous communications for devices on this bus without any bus contention. Some of devices here are the 32 Core CPU, the 4 channel DMA (Direct Memory Access Controller) with 2KB RAM, and USB 2.0-compliant Full-speed OTG ("On The Go") controller. Other components on this bus are 128K FLASH, 32K RAM, interrupt controller, all the digital ports, and a peripheral bridge (a connection to all the on –chip peripherals). It is interesting to note that digital ports reside on the high speed bus. This means they can be toggled (on/off) at the 40 MHz rate ---able to generate digital signals of up to 20MHz if needed. With the use of the DMA, this port (or any other peripheral), can directly access memory without CPU (software) intervention for data transfers.

The other bus is the peripheral bus. This bus does not need to run as fast as the previous bus and can be slowed down to a factor of 1 to 8 (peripheral clock rates are not required to be run as high as 40 MHz). The peripheral bus connects to all the on-chip peripherals. The PIC32MX2 also has an impressive and very extensive set of on-chip peripherals. Features include:

- ADC configurable with up to 9 analog channels
- 10 bit at 1.1Msps sample rate
- Three built-in analog comparators
- Parallel Master port with address capability
- A real time clock calendar chip (RTCC)
- Five 16-bit timers with reconfiguration to 2- 32 bit timer 1- 16 bit timer
- Five input capture –automatic pulse capture and measurement
- Five Output Compare / PWM modules –automatic pulse generation
- Two UART (Universal Asynchronous Receiver Transmitter) with address capability

- Two SPI (Synchronous Peripheral Interface) and I2S Audio Interface
- Two I2C (Inter Circuit Communication) interface
- CMTU (Charge Time Measurement Unit)

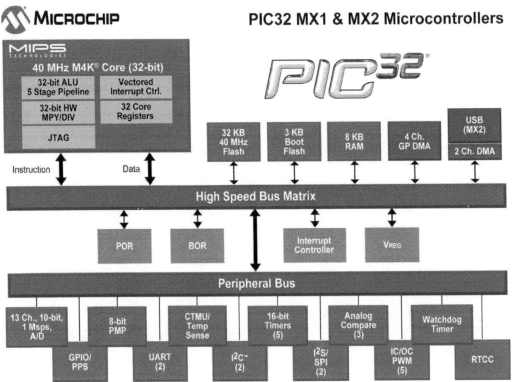

Figure 1-23: PIC32MX250F128B overview

The PIC32MX250F128B pin layout is shown along with each individual pin assignment. The PIC32 as shown is the table is a +3.3V part, however it can accept +5V logic level inputs on certain pins (see Figure 1-25 for +5V tolerant pins) without damage. Note that each pin has multiple assignment possibilities. Some of these pins are fixed function but others can be configured as analog, digital, digital Input with change detection, or assigned to a peripheral. The assignment depends upon how you want to use the device. The Microstick II provides a subset of these pins for prototyping use. Figure 1-24 shows which pins are available.

We listed the peripheral capabilities of the PIC32MX2 earlier. Note that the total peripheral content is large and, in fact, with these 28 pin PIC32MX2 devices, there are more internal peripherals then there are available pins. To counter this shortfall Microchip came up with a novel solution called Peripheral Pin Select (PPS). The PIC32MX250F128B supports PPS with 19 remappable pins as shown. With these pins (designate with "RPxx" for reprogrammable pin) you can configure which peripherals can appear on which pins, it is very much like programming your own pin layouts.

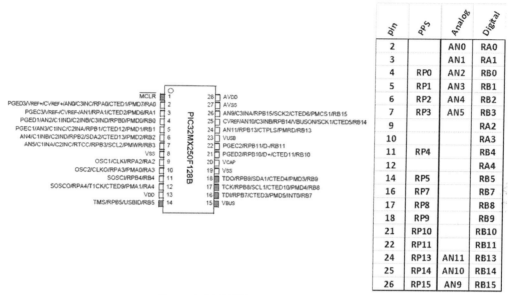

Figure 1-24: PIC32MX pin descriptions

pin	PPS	Analog	Digital
2		AN0	RA0
3		AN1	RA1
4	RP0	AN2	RB0
5	RP1	AN3	RB1
6	RP2	AN4	RB2
7	RP3	AN5	RB3
9			RA2
10			RA3
11	RP4		RB4
12			RA4
14	RP5		RB5
16	RP7		RB7
17	RP8		RB8
18	RP9		RB9
21	RP10		RB10
22	RP11		RB11
24	RP13	AN11	RB13
25	RP14	AN10	RB14
26	RP15	AN9	RB15

The PIC32MX2 microcontroller also has configurable clock options that are important in supporting chip level power management during idle, sleep and doze modes of operation. These modes incorporate fast wake-up, and switching between clock sources in real time. This feature allows you to switch to 40MIPS operation only when you need it, thereby conserving power, an important feature for battery operations.

As a final note with the PIC32MX2, you not only program its flash memory with object code, but you also program its configuration setting by programming its internal "configuration word". The configuration word setting can be set using the MPLAB X GUI or integrated in the source code as complier directives. The configuration word itself resides in flash memory but it is a reserved location and not considered as part of the code area. The configuration is captured in project file System.h.

Let's now get into the Microstick II operational details. To reiterate, the Microstick II is designed to provide users with an easy to use, economical programmer/debugger for 32 bit and 16-bit Digital Signal Controllers and Microcontrollers and is fully integrated with Microchip's MPLAB X Integrated Development Environment.

Microstick II supports all the essential features for programming/debugging. It can be used stand-alone or plugged into a prototyping board. Once you are satisfied with your design and feel that all the bugs are fixed, you take the microcontroller out of debug mode, and then program the device for standalone operation (release mode). In release mode MPLAB is no longer required. The only requirement here is to retain the USB connection to the Microstick II to retain power to the microcontroller.

A final option is to remove the microcontroller from the Microstick II and plug the microcontroller directly into the prototype without the Microstick II. For this you need to furnish your standalone microcontroller with power, ground, and Master Clear. We will cover this process in the some of the chapters to follow.

Let's discuss the detailed operation of the Microstick II hardware starting with a block diagram of the device (see Figure 1-25).

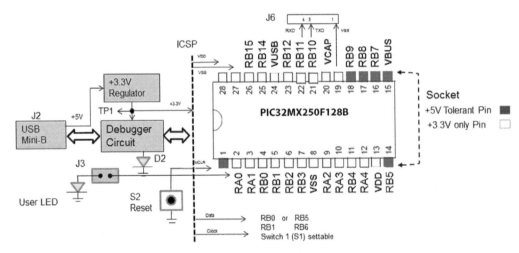

Microstick II PIC32MX2 Block Diagram

Figure 1-25: Microstick II Block Diagram

Integral to the Microstick II programmer/debugger is the target microcontroller. This microcontroller resides in a 28 pin socket that sits on the top side of the Microstick II board. There is another 28 pin socket that resides on the bottom of

Beginner's Guide to Programming the PIC32

the Microstick II just below the microcontroller. This socket is used to connect the Microstick II with a solderless bread board. This allows the Microstick II to perform as the microcontroller in your prototype.

Here the pins that are not involved in supplying power, master clear or debug/programming, are made available for prototyping. In total the Microstick II really provides about 19 pins of its 21 pins to the user. Note that some of these available pins are +5V tolerant and some are not. The +5V tolerant feature allows pins to accept +5VDC as a logic level to the pin without damage even though the part is +3.3VDC. Those that are not marked +5V tolerant can only accept +3.3VDC levels, exceeding this level will result in damage to that pin's electronic functions.

When you connect the solderless breadboard to the Microstick II using bottom side socket you need to plug in dual 14 pin male .100" headers to each side of the bottom 28 pin socket.

The Microstick II itself supplies +3.3VDC power to the PIC32 and all other on-board electronics by regulating the available +5VDC from J2 mini-B device USB connector (once it is plugged into an available PC USB port).
The Microstick II has its own on board debugger/programmer electronics that communicates with the MPLAB. This electronics resides on the bottom side of the board. The communication with MPLAB and the Microstick II is over USB and with this communication that MPLAB controls debug and programming. The on board programming and debug interface between Microstick II and the microcontroller uses Microchip's proprietary five pin interface known as the "In Circuit Serial Programming Interface" or ICSP. ICSP requires two pins on the target microcontroller, one for data and the other clock. On the PIC32MX2 these are pins 4 and 5 or optionally pins 14 and 15 (selectable by side switch S1). In addition, as noted earlier, ICSP also uses Master Clear pin 1 of the PIC32MX2. This pin is shared with an external pushbutton (labeled Reset on the top of the board) to allow the user to perform manual reset to their designs when required.

An LED for status of the debug circuit is visible from the top of the board. If the Microstick II debug circuitry is communicating with MPLAB then the LED remains on steady. A final LED is available for the user for prototyping. It is connected to pin 2 of PIC32MX2, and selected via jumper J3, more on this later.

Review of MPLAB-X Key features:

- Integrated development environment for Microchip Microcontrollers

- Uses Integrated Project Manager to organize and maintain project files and user's workspace
- Uses syntactic source code editors supporting both 'C' and assembly languages
- Single button build and program functions
- Support programming and debug of microcontroller using external devices like Microstick II
- Tool bar for run, pause, single step during debug operation.

Review of Microstick II Key Features:

- Low Cost
- Integrated USB programmer / debugger – No external debugger required
- USB Powered – Ease of use, No external power required
- Socketed PIC32MX2/dsPIC/PIC24 – Flexible, Easy device replacement
- 0.025" Pin headers – Enables plug-in to Breadboard with room for jumper wires
- Easy access to all device signals for probing
- Small size - Smaller than a stick of gum at 20 x76mm – Easily Portable
- On board debug LED, Utility LED and Reset Switch

Exercises:

1. What are some of key tools supported by the MPLAB-X environment?
2. What is meant by ICSP? And how is it used by the Microstick II?
3. Gain familiarity with Microchip's 32 bit family of products and target applications. Go to and review the materials at Microchip web site http://www.microchip.com/pagehandler/en-us/family/32bit/
   a. What are some applications?
   b. What free libraries are available?
   c. What are the more innovative peripherals listed for this family? And how does the PIC32 support them?
4. Gain familiarity with the PIC32 application notes. Go to URL http://www.microchip.com/stellent/idcplg?IdcService=SS_GET_PAGE&nodeId=1445 and select PIC32 as processor and do search. How many applications notes are currently listed?

5. How many +5V tolerant pins does the PIC32MX2 support?
6. What is PPS feature of the PIC32? How is it used?
7. What is a meant by the microcontroller's configuration word? How is it used? How can it be set?
8. Exploring MPLAB-X
   a. Open Demo project FLASH_LED
   b. Close all windows.
      i. Close project view
      ii. Close source code views
      iii. Close output window
   c. Open all windows (just basic IDE panel and tool bar should be present)
      i. On IDE panel select view and from pull down select project—the project window should be opened and project tree visible
      ii. Double click in project window on Flash_LED (this is the main function) – the source code should be open and source visible.
      iii. Click build button on tool bar –notice that output window with build tab is automatically opened and results from the build are shown.
      iv. Connect the Microstick II and notice the output window. It changes tab from build to debug and automatically recognizes the connection.
      v. Click run, then pause, then single step and follow the green arrow in the source window indicating where execution has stop and what the next line of code execution is. Now hit reset and notice arrow position, hit single step and notice the green arrow is now place at the beginning of the source code. Hit run and the Microstick II should be blinking the user LED.

Chapter 2 – Driving an LED Display

In this chapter we will develop prototype hardware and software for an LED display. We will be using the installed development system created in Chapter 1. For our first experiment we will start with just one LED. This is reminiscent of the 'C' program "Hello Word!" Here the programmer strives to write their first application program outputting a hello world message to the console. With our microcontroller there is no console so we have to just settle to blink a LED as a type of "Hello World". Here we will walk through hardware and software builds, and do the software debug using simulation and the finally do hardware debugging using Microstick II. In the second exercise we expand the hardware prototype capability by increasing the number of LEDs to a total of ten and enhance the software to perform a binary counting display.

PIC32MX Programmable Port overview

The PIC32MX has user programmable pins. These pins can be configured as digital inputs, digital outputs, analog, or assigned to a specific PIC32MX peripheral in/out. A block diagram (see Figure 2-1) highlights the inner organization of an individual PIC32MX pin to show just how this works. The diagram depicts only the peripheral and digital functionality of the pin, analog is not shown.

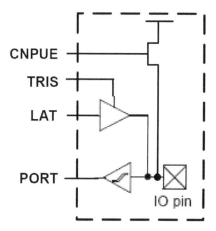

Figure 2-1: Block diagram of PIC32MX pin electronics

A pin's output can originate from either the peripheral side or the digital side. Both have their own independent data paths and enabling signals through an output mux which determines which side becomes the pin's output. A peripheral must be enabled for its output to be used. To enable a pin for digital output the pin's TRIS signal must be cleared. Only one side (peripheral or digital) can output at any one time, and if both are enabled, the peripheral with higher priority wins! On the input side no mux exists and so input signals connect to both sides. In this chapter we will use digital output only.

Every pin has a corresponding TRIS control bit to set the digital direction of the pin. A zero makes the pin an output; a one makes the pin an input. Each pin also has a PORT bit that allows reading or writing to the pin. For our PIC32MX part we have two PORT registers PORTA and PORTB that are both 16 bits wide. However, because of the 28 DIP package size not all of these 16 PORT bits are available as pins.

PORTA allows for read/write of pins RA0, RA1, RA3, RA4, these pins correspond to bits 0, 1, 3, 4 of PORTA (the rest of the bits are unassigned).

PORTB allows for read/write of pins RB0, RB1, RB2, RB3, RB4, RB5, RB6, RB7, RB8, RB9, RB10, RB11, RB12, RB13, RB14, RB15 again each of these pins corresponds to a bit position in the register.

As mentioned earlier, for each PORT we have a TRIS. So in this device we have a TRISA and a TRISB. Both PORT and TRIS exist in the PIC32MX as "special function registers" or SFR for short. They are 16 bit memory locations in the PIC32MX dedicated to digital pin operation. The pin bit positions in a PORT are identical to those in its TRIS. For example if RB2 is to be an output then write a zero to bit position 2 in the TRISB register and then set that pin's output value by writing a one (for +3.3VDC output) or zero (ground) to the corresponding bit position in PORTB.

Another SFR that is used for digital I/O is called LAT or latch. It works in a similar way to PORT, however with LAT we can latch output or input rather than directly reading or writing using PORT. We have both a LATA and LATB; again both are 16 bits.

Another SFR used for each I/O pin is weak pull-up and a weak pull-down control. The pull-ups act as a current source or sink source connected to the pin, and eliminate the need for external resistors when push-button or keypad devices are connected. The pull-ups and pull-downs are enabled separately using SFR(s) CNPUx and CNPDx registers, which contain the control bits for each of the pins. Setting any of the control bits enables the weak pull-ups and/or pull-downs for the corresponding pins. PORTB for example would use CNPUB and CNPDB.

Beginner's Guide to Programming the PIC32

Another important SFR feature set for the PIC32MX is the ability to do single instruction cycle bit manipulation. Every I/O module register has a corresponding CLR (clear), SET (set) and INV (invert) SFR register set designed to provide fast atomic bit manipulations. As the name of the register implies, a value written to a SET, CLR or INV register effectively performs the implied operation, but only on the corresponding base register and only bits specified as '1' are modified. Bits specified as '0' are not modified. Reading SET, CLR and INV registers returns undefined values. To see the effects of a write operation to a SET, CLR or INV register, the base register must be read.

There are various software techniques to access special function registers, like TRIS, PORT, and LAT in C code. We will discuss this some more when we talk about the software. For instance reading/writing a 16 bit word to the TRSID can use the register's CLEAR address (e.g. TRISDCLR), the register's SET address (e.g. TRISDSET), the register's INVERT address (e.g. TRISDINV), or using a 'C' language structure to access bit 'x' on all PORT SFR (example of PORT B):

- TRISBbits.TRISBx
- LATBbits.LATBx
- PORTBbits.RBx
- CNPUBbits.CNPUBx (pull-up enable)

PIC32MX IO using Peripheral Library

A very useful feature of the PIC32MX C compiler environment is the Microchip's peripheral library. With this library you can simplify all the nuances of ports and peripheral control without directly writing to any SFR .The PIC32 Peripheral Library Functions can control the PIC32MX PORTS, initialize the Core Timer or handle interrupts to name just a few. Here are a few examples for controlling PORTB I/O pins. You can use bit masks or literals as shown.

- mPORTBSetPinsDigitalOut(BIT_0 | BIT_1 | BIT_2);
- mPORTBSetPinsDigitalOut(7);
- mPORTBSetPinsDigitalIn(BIT_6 | BIT_7 | BIT_13);
- mPORTBSetPinsDigitalIn(0x20C0);
- mPORTBClearBits(BIT_0);
- mPORTBClearBits(1);
- mPORTBToggleBits(BIT_7);
- mPORTBToggleBits(0x80);
- mPORTBReadBits(BIT_2);
- mPORTBReadBits(4);
- mPORTBWrite(0x00FF);

- value = mPORTBRead();

A Help page is available to assist you with the library and was automatically installed as part of the compiler. Navigate to the location shown in Figure 2-2 and double click PIC32MX peripheral library. The library comes up with directories to all the PIC32 peripheral function calls along with detailed examples. We will show the use of library in some of the exercises to follow. The real power of this library comes into play not so much with the port configuration (which is fairly straightforward) but when configuring the more complex peripherals like I2C, SPI, and DMA which have lots of configuration SFR(s) associated with them.

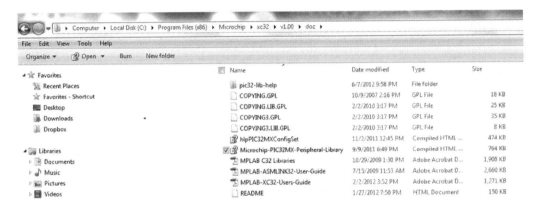

Figure 2-2: Navigating to Peripheral Library Help

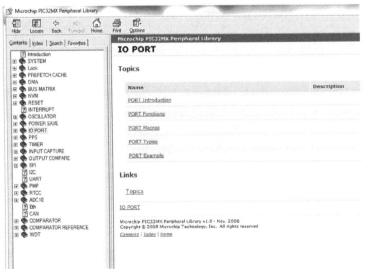

Figure 2-3: Microchip Peripheral Library Help

Experiment 1- Hardware

For Experiment 1 we want to connect a single LED to bit 5 of PORTB or RB5 and set RB5 to be a digital output. RB5 is pin number 14 on the PIC32MX. The device is a 28 pin DIP, so there are 14 pins on each side of the chip. Looking down on the chip there is a notch in the package to indicate pin 1, all the pins are numbered 1 to 28 and go counter clockwise on the device. This would make pin 14 or RB5 the last pin on the bottom right hand side of the chip.

A schematic for Experiment 1 is shown in Figure 2-5. The PIC32MX digital output can drive up to 25 milliamps on a digital output. We don't need all this current to turn on an LED (3 milliamps is more than enough), so we use a current limiting resistor (1K) in series with the LED. The circuit is basically a series circuit where RB5 is tied to one side of the resistor, the other side of the resistor is connected to LED anode, and finally the LED cathode is tied to ground. In this way when we make RB5 =1 the LED will light up, if we make RB5 =0 the LED will turn off. Make sure that the ground potential for your prototype circuit is the same ground reference used by the Microstick II, so connect the Solderless breadboard ground to pin 8 of Microstick II.

An assembled circuit is shown in Figure 2-4. Notice that ground is picked up from pin 8 and connected to a common bus for ground on the prototype board. This same ground is used for the LED cathode. The cathode must be grounded for the LED to work. You need to be able to figure out anode from cathode with the LED, to be able to hook it up correctly. The cathode is typically the shorter lead on the device is also marked by a flat edge on the part.

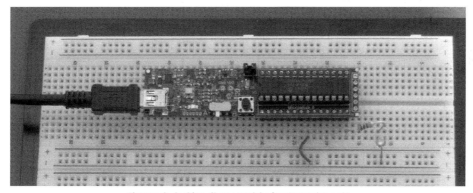

Figure 2-4: Circuit assembly for Experiment1

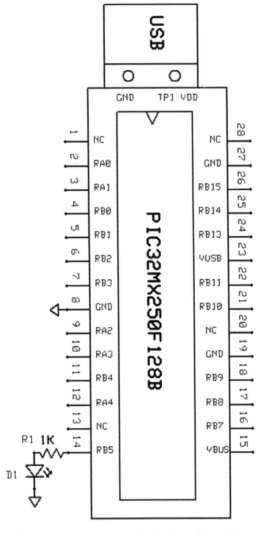

Figure 2-5: Schematic for Experiment 1

Experiment 1- The Software

Ok, the hardware is in place, lets discuss the software. We show both library and non-library versions. First look will be the non-library version where we will write directly to the SFR ourselves. Let's take time to review the MPLAB setup and software for Chapter 2's exercise 1. Open MPLAB-X and then open project by navigate to Chapter 2 code folder, exercise 1. In this folder double click the MPLAB X project icon FlashLed. You should see the following without the highlights.

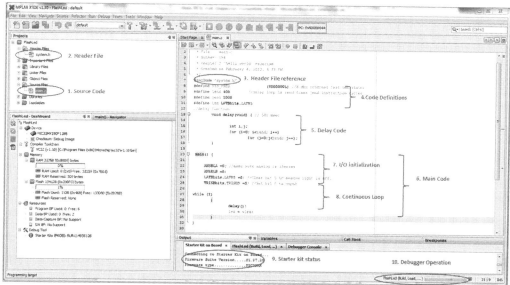

Figure 2-6: Survey of Experiment1 Workspace

The highlights cover a number of important areas within this project's workspace. Let's discuss them one at a time.

1. Source Code- Project and Editor View

    a) The source code is really the principal design. It is a .C file extension that is ANSI C compliant with the only exceptions being operations to Microchip processor specific function register (SFR) settings, and controls
    b) You can view the contents of the source code from the project view by double clicking the source code listing in the project view. –The source code will appear in an Editor Window.
    c) The Editor window will open with the contents of source code with a color code based upon language syntax. For example comments are purple, C language elements are dark blue. The color scheme, and fonts are all configurable under Tools->Options-> Fonts/Colors.
    d) In The Editor Window you can modify the source code or add new code
    e) The source code has several major components: generic header file, function code and finally the Main function

2. System.h Project View

Beginner's Guide to Programming the PIC32 — Page 47

a. It is known as a header file (because it has an .H extension). You can view the contents of the System.h from the project view by double clicking the source code listing in the project view. The System.h code will appear in an Editor Window, here you can view, modify or add new code.
b. It contains two major parts the references part and library references that are developed and supplied by Microchip, and the fuse configuration for how we want to use the part. We will discuss both. The part reference contains internal 'C' code structure SFR definitions that are needed in supporting reading/writing to the PIC32MX microcontroller peripherals and I/O in an ANSI C compliant language environment. The library reference allows us to access and use Microchip peripheral library functions. The parts and library reference is achieved with the following line of code in System.h.

```
#include <plib.h>
```

c. The actual files exist outside our project. Using the <> syntax, it references the Microchip program file directory on your C: drive.

3. System.h Editor View

A code snippet from the System.h file is shown using Editor Window. All constituent parts are shown in Figure 2.7. Note the location of the #include <plib.h>. Next there are four unique logical grouping of PIC32MX fuse configurations DEVCFG3, DEVCFG2, DEVCFG1, and DEVCFG0. To set a configuration we use a 'C' code complier directive #pragma followed by config followed by the particular fuse setting. The configuration settings that we will be using for our exercise is as follows:

```
// DEVCFG3
#pragma config IOL1WAY = ON         // Peripheral Pin Select Configuration (Allow only one reconfiguration)
// DEVCFG2
#pragma config UPLLEN = OFF         // USB PLL Enable (Disabled and Bypassed)
// DEVCFG1
#pragma config FNOSC = FRC          // Oscillator Selection Bits Fast RC 8MHZ
#pragma config FSOSCEN = OFF        // Secondary Oscillator Enable (Disabled)
#pragma config IESO = OFF           // Internal/External Switch Over (Disabled)
#pragma config POSCMOD = OFF        // Primary Oscillator Configuration (Primary OSC disabled)
#pragma config OSCIOFNC = OFF       // CLKO Output Signal Active on the OSCO Pin (Disabled)
#pragma config FPBDIV = DIV_1       // Peripheral Clock Divisor 1
#pragma config FCKSM = CSDCMD       // Clock Switching and Monitor Selection disabled
#pragma config FWDTEN = OFF         // Watchdog Timer Enable (WDT Disabled)
// DEVCFG0
```

```
#pragma config JTAGEN = OFF        // JTAG Enable (JTAG Disabled)
#pragma config ICESEL = ICS_PGx1   // ICE/ICD Channel Select (Communicate on PGEC1/PGED1)
#pragma config PWP = OFF           // Program Flash Write Protect (Disable)
#pragma config BWP = OFF           // Boot Flash Write Protect bit (Protection Disabled)
#pragma config CP = OFF            // Code protection is not enabled.
```

Here a synposis of our configuration:

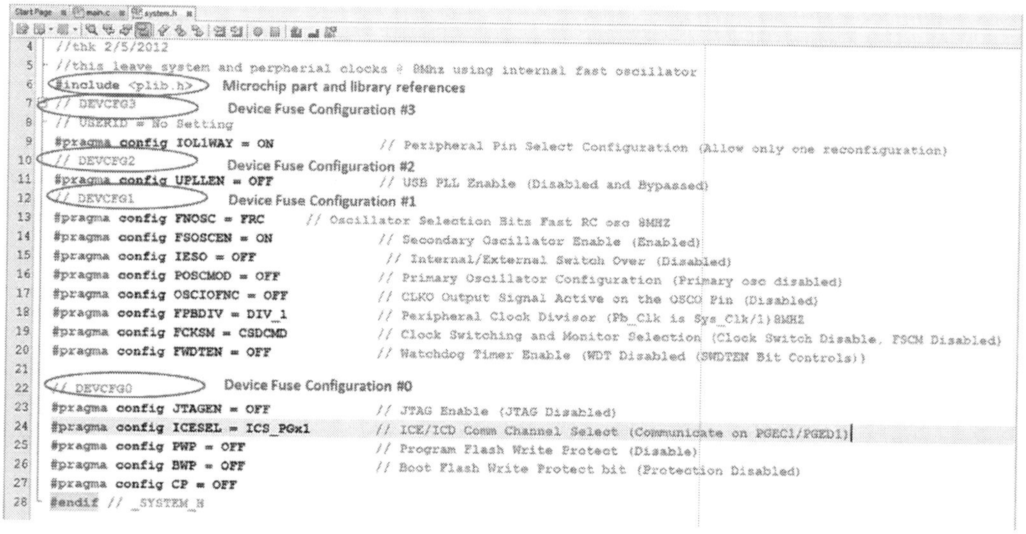

```
//thk 2/5/2012
//this leave system and perpherial clocks @ 8Mhz using internal fast oscillator
#include <plib.h>        Microchip part and library references
// DEVCFG3               Device Fuse Configuration #3
// USERID = No Setting
#pragma config IOL1WAY = ON          // Peripheral Pin Select Configuration (Allow only one reconfiguration)
// DEVCFG2               Device Fuse Configuration #2
#pragma config UPLLEN = OFF          // USB PLL Enable (Disabled and Bypassed)
// DEVCFG1               Device Fuse Configuration #1
#pragma config FNOSC = FRC           // Oscillator Selection Bits Fast RC osc 8MHZ
#pragma config FSOSCEN = ON          // Secondary Oscillator Enable (Enabled)
#pragma config IESO = OFF            // Internal/External Switch Over (Disabled)
#pragma config POSCMOD = OFF         // Primary Oscillator Configuration (Primary osc disabled)
#pragma config OSCIOFNC = OFF        // CLKO Output Signal Active on the OSCO Pin (Disabled)
#pragma config FPBDIV = DIV_1        // Peripheral Clock Divisor (Pb_Clk is Sys_Clk/1)8MHZ
#pragma config FCKSM = CSDCMD        // Clock Switching and Monitor Selection (Clock Switch Disable, FSCM Disabled)
#pragma config FWDTEN = OFF          // Watchdog Timer Enable (WDT Disabled (SWDTEN Bit Controls))

// DEVCFG0               Device Fuse Configuration #0
#pragma config JTAGEN = OFF          // JTAG Enable (JTAG Disabled)
#pragma config ICESEL = ICS_PGx1     // ICE/ICD Comm Channel Select (Communicate on PGEC1/PGED1)
#pragma config PWP = OFF             // Program Flash Write Protect (Disable)
#pragma config BWP = OFF             // Boot Flash Write Protect bit (Protection Disabled)
#pragma config CP = OFF
#endif // _SYSTEM_H
```

Figure 2.7: Editor View of System.h

4. Configuration

   a) DEVCFG3 allows only one programmable pin configuration during program execution. We talk more about this in a later chapter, but the PIC32MX allows its pins to be custom mapped by the user to its internal perpherial as per user requirements. It is like laying out your own chip pinout. For reliability we limit that through DEVCFG1 to once in the program application.

   b) DEVCFG2 controls the clock for the PIC32MX on-chip USB perpherial . Since we are not using USB we are disabling the clock.

   c) DEVCFG1 controls the source of the PIC32MX System clock as well as the rate of the perpherial bus clock (that is derived from the System Bus Clock). In this case we are using the PIC32MX internal 8MHZ RC oscillator as the system clock, because of this we are disabling the Primary and Secondary Clock systems. Because we are disabling all clock sources other then the internal 8MHZ RC we shutdown any automatic capability for switching between external and internal clock sources.

The PIC32MX has the ability in the event of a primary clock failure to switch automatically to a secondary clock source. With the Primary clock shut down we are using one of its OSC inputs to the chip to be reassigned as a digital pin.
Finally we are disabling the Watchdog Timer. The Watchdog timer is an internal hardware based timer inside the PIC32MX and is used as a failsafe feature, in that if it runs to completion and overflows it will automatically reset the microcontroller. The designer configures the watchdog period and ensures that the microcontroller resets the timer before this period is reached. By definition if the microcontroller can do this it is functioning normally. If there is an anomaly in operation and the microcontroller fails to reset the watchdog timer, the timer will reset the microcontroller operations, forcing it to initialize all operations from the beginning, restoring normal operation.

For prototyping activities it is just not required, and more of a burden in the early embedded design cycle. It is really intended for use in a final product.

d) DEVCFG0 turns off JTAG programming capability and turns on the Microchip in Circuit Serial Programmer (ICSP) that is built-into the Microstick and then configures ICSP to use the right clock and data lines as needed by the Microstick II. Finally DEVCFG0 disables all BOOT/Program FLASH write protection (not required for prototyping) and turns off Code protection for the same reasons.

e) By placing all critical configuration and library/part reference data in one file it can be simply referenced in your Main code and any configuration changes are limited to only one file in your project.

5. Code definitions

   a) These are just C language define statements that allow naming substitutions for complex terms to simpler terms. This makes the code easier to read. For instance we have an LED (in series with a 1k resistor to ground at PIC32MX250128B Pin 14 or RB5.) We want to write to the LATB register at bit position 5 to either turn on or off the LED. This specific bit address in the Special Function Register of the PIC32MX is LATBbits.LATB5. Rather than continuing writing this out with simply redefine it we the simpler term "led" using the 'C' #define compiler directive and the SFR 'C' structure for LATB

   ```
   #define led LATBbits.LATB5
   ```

   b) Here are three possible ways of accessing SFR LATB bit, we will show them here for RB5 TRIS and LAT, setting RB5 to be an output and then driving it high.

   i. C Structure format for RB5 TRIS and LAT
      1. TRISBbits.TRISB5 =0 ; //set just RB5 to be an output
      2. LATBbits.LATB5 =1 ; // drive just RB5 high

   ii. Word entire word to SFR
      1. TRISB = 0xFFDF ; //set RB5 to be and output (all others inputs)
      2. LATB = 0x0020 ; //drive RB5 high all others low

   c) The other defines are limits used by the delay function and for system clock rate. We set these values here in one place in our Main function for use by other parts of the application code. A change in any of these defines is then limited to one location in the code.

6. Delay function

a) In order to blink the LED we need some delay time for turning it on and turning it off in order to allow it to persist in our vision as a "blink". This function provides this. It is configured with a fixed delay, that idles and does nothing else but consume CPU cycle time for about ½ second. The delay is configured as a loop within a loop. The loop times are established by **iend** and **jend** values. The total delay time is approximately **iend** time's **jend** cycles or 400 x 1000 or 400,000 cycles to achieve 500 millisecond delays.

```
void delay(void) { // 501 msec
    int i,j;
    for (i=0; i<iend; i++)
        for (j=0; j<jend; j++);
}
```

Figure 2.8: Delay Function

7. Main code

a) Every C program must have a single function called "main". This is where execution begins for the user, only after the compiler sets the configuration word, initializes the processor stack pointer, and initializes all the variables. Only one main function can exist in a C language application at a time. As we will see the main function is often segmented with an initialization section and a continuous loop section. Once initialization is complete the microcontroller enters the continuous ("infinite") loop for its operations.

This is marked by a While (1) { ……}. The microcontroller normally never leaves this part of main unless a power reset or watchdog timeout occurs.

```
main() {

    ANSELA =0; //make sure analog is cleared
    ANSELB =0;
    LATBbits.LATB5 =0; //Clear bit 5 to ensure light is off.
    TRISBbits.TRISB5 =0; //Set bit 5 to ouput

while (1)
        {
                delay();
                led = ~led;
        }
}
```

Figure 2.9: Main Source Code Experiment 1

b) Normally PC based C applications will run on and then return back to the PC operating system once the main function is finished. For microcontroller applications there is no Operating system so if main terminates, the complier resets the processor by starting execution from the beginning.

c) I/O initialization

   i. This is within the main function, as mentioned earlier, and it is the segment of main that is dedicated to initializing all microcontroller ports and peripherals (SFR) before the application can begin to use them. Initialization usually happens once in main before it enters into a continuous operating loop. Four special function registers (SFR) are involved for our application. ANSELA and ANSELB are used and configured to turn off all analog associated capability to those pins that have analog content. LATB is a latch output setting for all pins associated with PORTB pin outputs, in this case all latched settings are zeroed —nothing will appear as a high level on output to start. TRISB is the directional setting for PORTB that in this case makes them all outputs.

```
ANSELA =0; //make sure analog is cleared
ANSELB =0;
LATBbits.LATB5 =0; //Clear bit 5 to ensure light is off.
TRISBbits.TRISB5 =0; //Set bit 5 to ouput
```

Figure 2-10: Code Initialization Experiment 1

d) Continuous loop

    i. This is the primary working segment of microcontroller applications. The continuous loop is where all the work gets done. In our loop we invoke the delay function and after this we "toggle" the current value of LED (which is defined as the LATB bit 5 or RB5 pin). By "toggling", if the current RB5 is value 1 then it is set to 0, if 0 then it is set to 1. In this way after every delay the LED connected to RB5 will blink.

```
while (1)
{
    delay();
    led = ~led;
}
```

Figure 2-11: Exercise 1 Main Loop

e) Microstick II Status

    i. IDE provides a Starter Kit Status tab under the task pane. Selecting this tab provides a window of the status of the Microstick II. Notice that other selection tabs are also available: compiler build, debugger console, and FlashLed compile build, more on these later.

f) Debugger Specifics

　　ii. This is an integral element of IDE system that highlights the debugger active status as a running progress bar.

A Library versus Non-Library Implementation

The approach for code implementation has been to use direct control of the PIC32MX SFR without the use of Microchip Peripheral Library. The use of library is optional for this application in that most of the SFR control is fairly simple. However, for completeness sake we also included a library based version of the code for your review. It works in an identical matter to the non-library version and can be accessed from Chapter 2 exercises as project LedFlash-library. Here's a snapshot of its Main code. All other code in the project is identical to the non-library version. Note that you can mix library and non-library functions within your code to get the job done. There are no constraints.

```c
#include "system.h"
#define SYS_FREQ            (8000000L) //8 Mhz internal fast oscillator
#define iend 400            //delay loop is iend times jend instruction cycles
#define jend 1000
//delay function
    void delay(void) { // 501 msec

        int i,j;
        for (i=0; i<iend; i++)
            for (j=0;j<jend; j++);
    }

main() {

    ANSELA =0; //make sure analog is cleared
    ANSELB =0;
    mPORTBClearBits(BIT_5);             //Clear bits to ensure light is off.
    mPORTBSetPinsDigitalOut(BIT_5);     //Set port as output
    while (1)
        {
            delay();
            PORTToggleBits(IOPORT_B,BIT_5);
        }
}
```

Figure 2-12: Library version of Exercise 1 Source

Experiment 1- How it Works

The application uses a single C source file (main.c) that is a main function that references a single generic header file (#include "System.h"). This header file determines the specific PIC32MX processor in use (PIC32MX250F128B), reference to the Microchip part libraries, and the required setting for the PIC32MX configuration fuses. The configuration settings are, for example, internal CPU clock selection of 8MHz, no external oscillator, no clock switch over, no watchdog operation, and use of RA3 as a digital/analog pin versus an external crystal pin input.

In the main function the application performs I/O initialization and then proceeds to enter a continuous loop to alternately turning on and then off an LED, using a delay between on and off settings. A ½ second delay is required to pause between settings so that we can see the LED "blink". The LED is wired from ground, through a series resistor, to pin 14 or RB5 of the PIC32MX250F128B. A flow diagram of this specific LED display application is shown along with the contents of the main source code file.

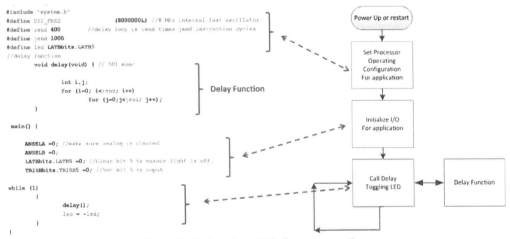

Figure 2-13: Experiment1 Software operations

Experiment 1 Execution

Let's go ahead and begin our debug. At this point we have a fairly good idea of what will happen during execution given that the prototype hardware is configured correctly, and the source code and processor configuration are in good shape. The PIC32MX will blink the LED connected to pin 14 RB5 of the Microstick II.

Let's verify the delay time and ensure that it is sufficient to yield a good blink rate. One quick way is to use the simulation tool "stopwatch" to measure this delay time. Under simulation there are some tool advantages to perform detailed analysis. Stopwatch is an example of this. Under "stopwatch" we can make actual time measurement. Let's start with setting up simulation and then use stopwatch to verify delay time before working with real time debugging and the Microstick II.

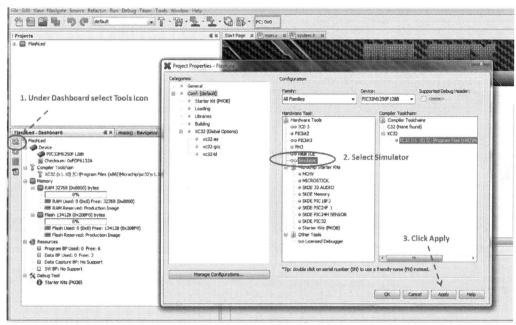

Figure 2-14: Selecting Simulator

You can verify that the simulator is the selected debug tool by refreshing and examining the Dashboard.

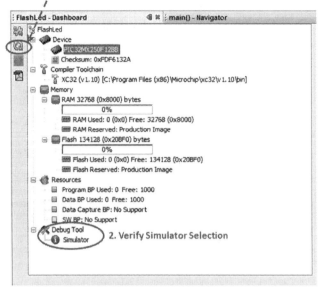

Figure 2-15: Verify Simulator Selection

Ok, at this point we should be good to go with the simulator, however, let's first make sure that the simulator setting for CPU clock is correct. As you may recall we set this to 8MHz in our system configuration, again under Dashboard select tools icon.

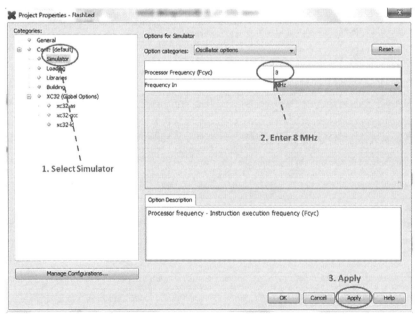

Figure 2-16 Setting Simulator CPU Clock

Beginner's Guide to Programming the PIC32

Let's set the breakpoint. Go to the source code. Position the cursor at the exact line where delay function is called in the main loop. Move the cursor to the grey left side line number column in the source and double click your mouse. A red breakpoint icon should appear aligned with source code that calls the delay.

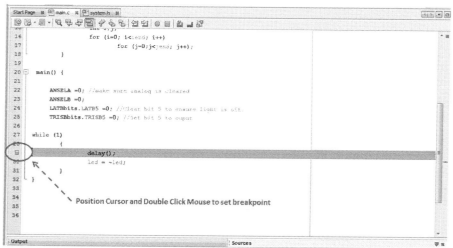

Figure 2-17 Setting a Breakpoint

Now let's select stopwatch tool for use with the simulator.
Go to Window -> Debugging-> Stopwatch.

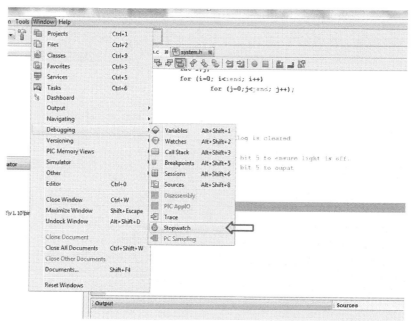

Figure 2-18 Initiating Stopwatches

Beginner's Guide to Programming the PIC32

Ok, we have the stopwatch and the breakpoint set correctly. We want to measure the delay between successive calls to the delay function. This should be good enough to determine blink rate. Hit Debug run and the processor will run to first breakpoint. Since we are running under simulation under 8MHZ, be patient, the IDE is running non-real time- fairly slow.

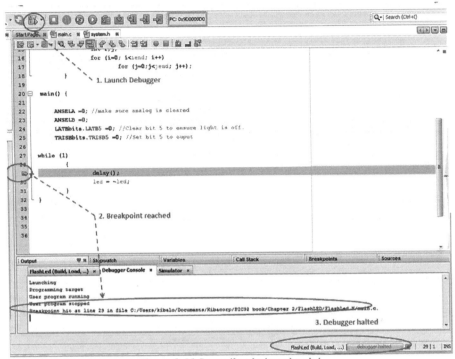

Figure 2-19 Execution to breakpoint

Note that with debugger launched the main toolbar has extended with the full debug tool suite visible. All of these tools are available during a debug simulation.

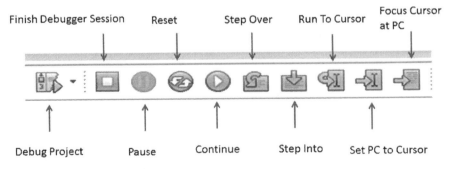

Figure 2-20 Main Debug Tool Suites

Beginner's Guide to Programming the PIC32

In the Output window select the stopwatch tab (see below). Note in lower right hand corner the simulation clock frequency setting of 8 MHz we set earlier. This first stopwatch reading includes execution time from the beginning of code to breakpoint; this is not what we are interested in. We are really interested in the time between breakpoints. Clear stopwatch by "clicking" Trash Can Icon and then click Continue icon in main tool bar to run without resetting.

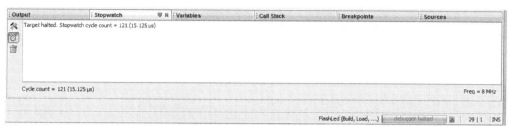

Figure 2-21: Using the Stopwatch initial break

We will then break again when the delay function breakpoint is reached a second time. Here we have captured an accurate time for the delay function.

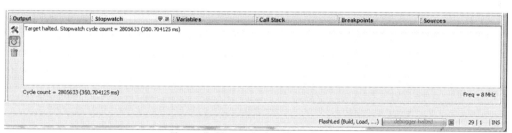

Figure 2-22: capturing Time Delay

The stop watch shows elapsed time of approximately 351 milliseconds. This means that the LED "on" time and the LED "off" time are both 351 milliseconds, meaning the LED has a blink rate at about .75 second, slow enough for us to see.

One other thing we can do under simulation is to check the toggle of RB5 using watch. Let's leave the breakpoint where it is and do this. We select "Window" from the tool bar, scroll down to watches and launch a watch tab under output.

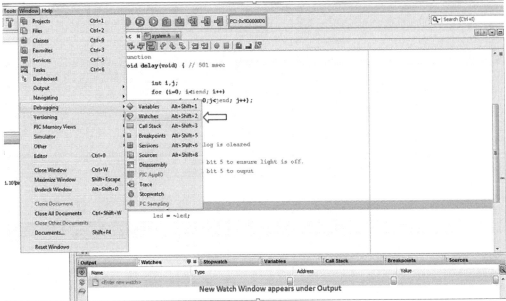

Figure 2-23: Selecting Watch

In this watch window we can view any named variable in our code or any SFR. Let's right chick on "Enter New Watch". A dialog box appears; here we wish to add LATB to our viewing.

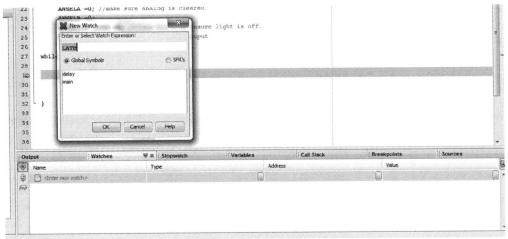

Figure 2-24: Setting Watch to LATB

Beginner's Guide to Programming the PIC32

We can view LATB as either word or bit view (by expanding word).

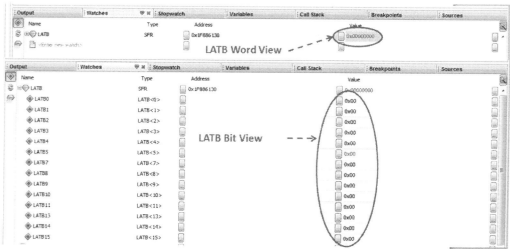

Figure 2-25 Different Watch View Options for LATB

By watching LATB between the breakpoints we should be able to see the RB5 turn on and off. Leaving watch open, with breakpoint set, click continue in the main tool bar. Note the value of LATB changes between breakpoints.

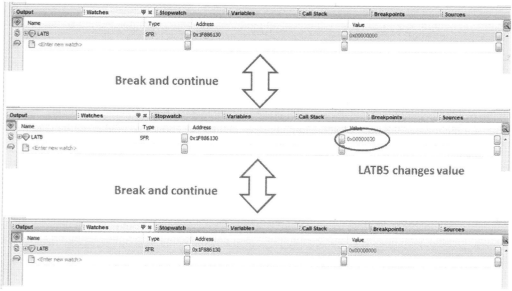

Figure 2-26: View Watch over number of breakpoints

Ok, we should now have confidence that our source code is working as designed. RB5 does in fact toggle and it toggles at a blink rate of every ½ second. All design elements check out. Let's now check it out for real with the prototype.

Let's reinitiate the debugging process with the Microstick II debugger. At this point we should have strong confidence that the software works. The only unknown at this point is the actual prototype hardware. Any "bumps in the road" should be attributable to our circuit wiring. Let's give it a try. It is pretty straight forward. First stop the debug process by selecting end debug session in main tool bar. The entire debug tool suite should disappear. From Dashboard move to Project Properties and re-select your debugger from simulator to starter kit.

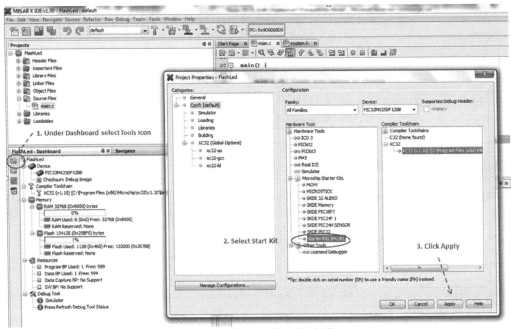

Figure 2-27 Selecting Start Kit

The MPLAB output window should go ahead and recognize the Microstick II is connected (assuming you have the Microstick II connected to the USB of your PC!). Refresh the Dashboard and you should see the following open tab in the Output area.

Figure 2-28 Insuring Start kit is activated.

The next to final step is removing the previous breakpoint (just double click on the breakpoint icon in the source window and it should disappear). Start a new debug session using the starter kit by hitting the Debug Project Icon button in the main tool bar. It should automatically build, download, and run the program. The debugger console under output should show this activity. The debugger activity progress bar should also be active.

Figure 2-29: Running Exercise 1 Application on Prototype

You should see your prototype blinking its LED. Any anomalies have to be associated with your hardware configuration –please check it carefully using the schematic and our assembly picture.

Congratulations! You have you first working prototype! If you are tired of the blinking LED you can stop at any time by clicking pause. To continue, continue or reset. Again, the entire debug tool suite should be active. For instance you can still set breakpoints and watch as needed. Stopwatch only works under simulation.

Experiment 2

Let's expand on what we have and add some new functionality. Here we use the same software structure as in experiment 1 but modify the software and add hardware to demonstrate an LED digital counter. Let's start with the hardware first. The schematic is shown contains 5 LEDs and series resistors to the prototype, also included is a photo of the new assembly.

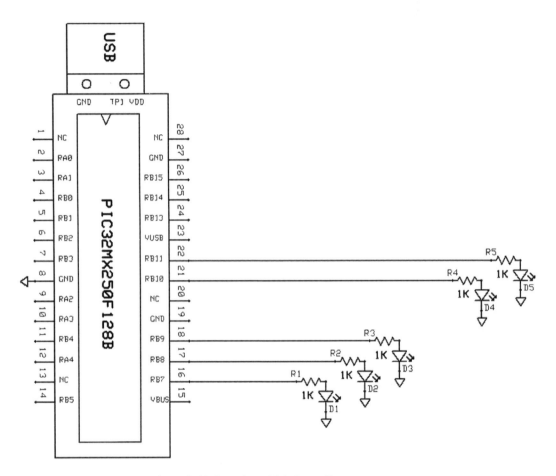

Figure 2-30: Experiment 2 Schematic

Figure 2-31: Experiment 2 Assembly

The objective here is to blink all the LEDs. We will use a binary counter. With a binary counter each successive LED in the LATB port will blink at a rate twice as less as the previous LED. Besides the hardware addition very little change needs to happen in software or the MPLAB workspace settings to make this happen. MPLAB X should be open form last exercise. Select File -> Open Project and then navigate to Chapter2 folder then "hello world with leds" folder. Click on HelloWorld.X MPLAB X icon. You should now have two projects viewable in the Project Pane.

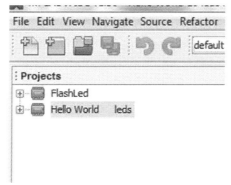
Figure 2-32: Two Project Workspace

Beginner's Guide to Programming the PIC32

MPLAB X has the ability to have many projects listed in the project window. We need to the select the new project as the main project. Right click on Hello World 10 LEDs and set as main project. It should appear bold in the project listing view against all the other projects. Dashboard automatically shows new project. You need to close out all the other code tabs (Main and System) and reopen them under the new project directories. It should appear as follows once you do a Build Main project.

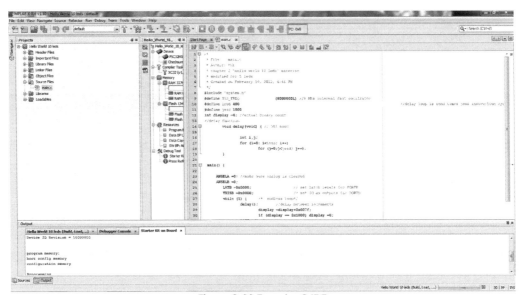

Figure 2-33 Exercise 2 IDE

There should be no compile errors. Let's examine the code changes before executing. Again you have the option of simulating, but it is really not necessary. Experiment 2 is a modification of Experiment1. Rather than simply toggling an LED (RB5 of LATB), we increment the entire LATB and display each available pin as a digital out. To achieve this we set up a new integer variable display. We then increment this display with each cycle through the continuous loop and then output the entire display value to LATB. Note that TRISD in initialization set all LATB pins to output.

We have available to us 5 LEDs on PORTB starting with RB11 to RB7 and designated as the "user LED". To increment we do so with a value of 0x2f. This treats RB5 as the least significant bit in our binary counter.

```
* File:    main.c
* Author:  thk
* chapter 2 "hello world 10 leds' excerise
* modified for 5 leds
* Created on February 10, 2012, 6:46 PM
*/
#include "system.h"
#define SYS_FREQ        (8000000L) //8 Mhz internal fast oscillator
#define iend 400        //delay loop is iend times jend instruction cycles
#define jend 1000
int display =0; //actual binary count
//delay function
        void delay(void) { // 501 msec

                int i,j;
                for (i=0; i<iend; i++)
                        for (j=0;j<jend; j++);
        }

main() {

    ANSELA =0; //make sure analog is cleared
    ANSELB =0;
      LATB =0x0000;              // set latch levels for PORTB
      TRISB =0x0000;             // set IO as outputs for PORTB
      while (1) {   /* endless loop*/
              delay();              //delay between increments
                display =display+0x007f;
                if (display == 0x1000) display =0;
              LATB =display;
        }

}
```

Figure 2-34: Experiment 2 Code

Click Debug project (make sure Microstick II is connected). The IDE will build, program and run the Microstick II. Watch binary counting light show! Any problems at this point must be attributed to hardware so double check with the provided schematic and assembly photos.

Review of PIC32MX250F128B Key features:

- Supports four user available digital pins for PORTA and 15 user available pins for PORTB in its 28 pin package
- ADSELA and ADSELB turns off any analog capacities that are inherent in pin port A and B hardware
- TRISA and TRISB control digital directional setting of the pin hardware. A

zero in the appropriate bit/pin position makes the pin an output; a one makes the pin an input. In default power up and for safety reasons all pins are set to input unless overridden by source code to other settings.

Review of Source Code Key Features:

- Source code is contains the main function.

- There is only one main function in a project

- The source code is segmented into features:
    - Use of a system reference header file
    - Setting configuration words
    - Setting I/O
    - Main loop –doing the required operation in a continuous fashion

- Main loop never should terminate in a microcontroller application, and if it does as a failsafe it resets the entire operation to start all over again.

- The plib.h header file resides in XC32 directory
    - It contains necessary macros and SFR definitions to both control fuse setting and configure SFR bits within a ANSI 'C'" language environment
    - There are three 'C' constructs to handle SFR
        - Whole port read and write directly
        - Structure format for individual bits
        - Short hand double underscore with Bit name

- Fuse configuration settings are done to the processor configuration word using predefined macros. These settings are no external oscillator, configuring an available oscillator pin (RA3) for digital use, selecting the internal RC oscillator running at 8MHZ as the microcontroller clock, turning off any clock automatic switchover (from external to internal) and finally turning off the watchdog timer.

- The Delay function is critical to the overall blink capability by providing sufficient time for the LED to stay on and off for visible persistence.

Review of MPLAB Debug Features:
- Simulation –Stopwatch to accurately time events, make sure to

- Set the internal microcontroller clock rate
- Set breakpoints
- Use watch to investigate key variable or SFR values at breakpoint
- Use Microstick II for real time debugging
- Source code can be removed and new code entered using Project View (if new code already exists in project space)

Exercise:
1. Explain DEVCFG3,2,1,0 and associated fuses.
2. Expand on Exercise 2
    d. Use same hardware configuration
    e. Modify source code to do an arbitrary LED light pattern
        i. Hint use a 'C' array with LATB patterns and index lookup to move individual LATB light patterns to LEDs during each loop delay
        ii. Solution: a source code file named 'array.c' existing within the current project space. Use techniques established in experiment 2 to introduce this new source. Examine this source for code example. Try modifying the light pattern for something different.

Chapter 3 – Reading an Input Switch

In this chapter we will develop prototype hardware and software to read a pushbutton. We will be using the installed development system from Chapter 1 and reuse a lot of the hardware setup from Chapter 2. In our first experiment will start with just one LED. In this case we will turn on the LED only when the pushbutton is depressed. In the second experiment we built a five stage binary counter display, but in this case we will use the pushbutton as the source of input clock to the counter. The trick here is to precisely count the number of pushbutton depression.

Although you may think this is straightforward, realize that the pushbutton is a mechanical device, with mechanical "chatter" associated with each depression and release. The processor is a very fast electronic device operating millions of instruction per second. It can actually "see" the mechanical chatter or "switch bounce" and with this, misinterpret the number of "true "pushbutton depressions. This results in an erroneous count to the user depressing the button. The picture in Figure 3-1 neatly shows the problem. Here the switch is pressed and then released. In these transition states the switch "chatters" between its on/off positions before settling. It is "switch bounce". The consequences of uncorrected switch bounce can range from being just annoying to catastrophic. For example, imagine advancing the TV channel, but instead of getting the next channel, the selection skips one or two. This is a situation a designer should strive to avoid.

Switch bounce has been a problem even with the earliest computers. The trick to reading the true switch state with a fast Microcontroller like the PIC32MX is to sense change and then wait a nominal settling time before making a switch state determination. This will be critical to obtaining an accurate count in Experiment 2. It is not an issue for Experiment 1. In Experiment 1 we turn on the LED when the switch remains pressed. Because the "chattering' is fast we don't see the individual LED on/off and our eye naturally integrates the LED on or off state.

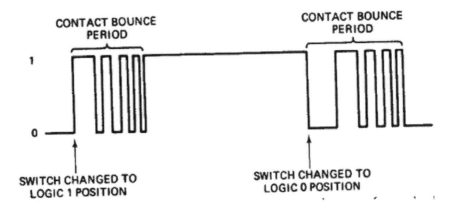

Figure 3-1: Mechanical Switch Bounce

Experiment 1- Hardware

As discussed in Chapter 2 PIC32MX has user programmable pins. These pins can be configured as digital inputs, digital outputs, analog, or assigned as specific peripheral. In this chapter we leverage the knowledge we gained in Chapter 2 and now use a digital input. As you may recall every pin has a corresponding TRIS control bit to set the digital direction of the pin. A zero makes the pin an output; a one makes the pin an input. Each pin also has a PORT bit that allows reading or writing to the pin and a LAT that works in a similar way to PORT, however with LAT we can latch an output or input rather than directly reading or writing using PORT. The LAT, PORT, and TRIS are special function registers or SFR. All these registers are 16 bits but with the reduced package 28 pin "skinny" DIP like the one are using with the Microstick II, not all of the SFRs bits are used.

There are a number of software techniques to access special function registers, like TRIS, PORT, and LAT in C code. We covered this in Chapter 2 and we will continue to expand on this some more when we talk about the software here. For Experiment 1 we connect a single LED to bit 5 of PORTB or RB5 (pin 14 of PIC32MX) and then set RB5 to be a digital output to drive an LED through a current limiting resistor. A pushbutton is connected to RB3 or pin 7 of the PIC32MX. This pin will be set to be a digital input.

Let's briefly discuss how the switch is connected. The switch is a mini-tactile pushbutton. It is called a SPST (single pole single throw) switch. The contacts are shown in Figure 3-2 as normally open with no electrical conduction through the switch. As you press the push button, contact is made across the switch. In using this switch we connect one end to ground and the other end to the RB3 digital input. When the switch is depressed the input will read ground or zero condition.

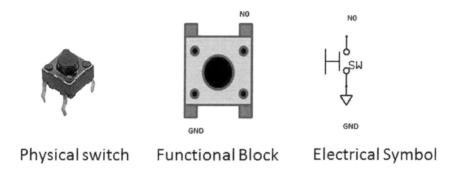

Figure 3-2: Switch Representation

When the switch is not pressed it will be open and the RB3 pin will float (not have any direct connection between ground or +3.3VDC). In this case, the condition as read by the microcontroller will be indeterminate. To remedy this situation, we need to tie +3.3VDC through a resistor to the NO terminal of the switch. In this matter when not pressed RB3 will read a one or high value, and when pressed, the switch will connect that point to ground and RB3 reading will be zero.

Rather than use an external resistor to make the circuit, the PIC32MX has a neat feature, the capability of configuring a pull up resistor internally to the pin through a 100K resistor. It is called a weak pull up in that it uses 100K so not much current is available, which is ok here. The pull up control is enabled by setting the appropriate bit within an SFR. In this case the SFR is CNPUB (Change Notification Pull Up). See Figure 3-3.

When you examine the pin out for the PIC32MX250F128B, pin 7 RB3, it shows that RB3 shares with CNPUB bit 3, so this is the control bit we want to set in the CNPUB SFR. By setting it, the RB3 pin will be pulled up through an internal 100K resistor to +3.3VDC, or the microcontroller supply, eliminating the need for an external pull-up resistor.

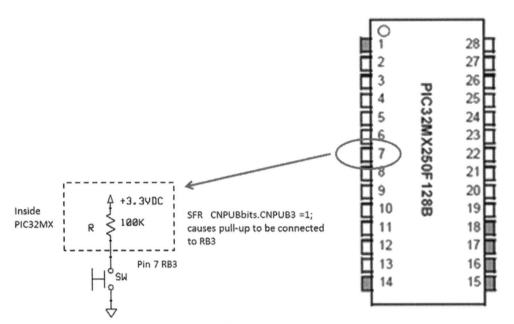

Figure 3-3: Setting Internal Pull-Up Resistor

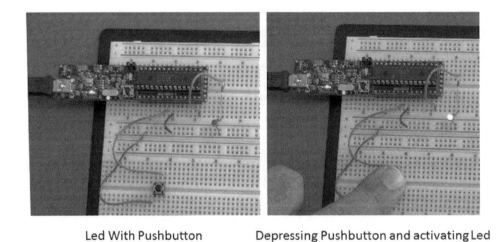

Led With Pushbutton   Depressing Pushbutton and activating Led

Figure 3-4: Experiment 1 Assembly

An assembled circuit is shown in Figure 3-4. Notice that ground is picked up from pin 8 and connected to a common bus for ground on the prototype board. This same ground is used for the LED cathode. The cathode must be grounded for the LED to work.

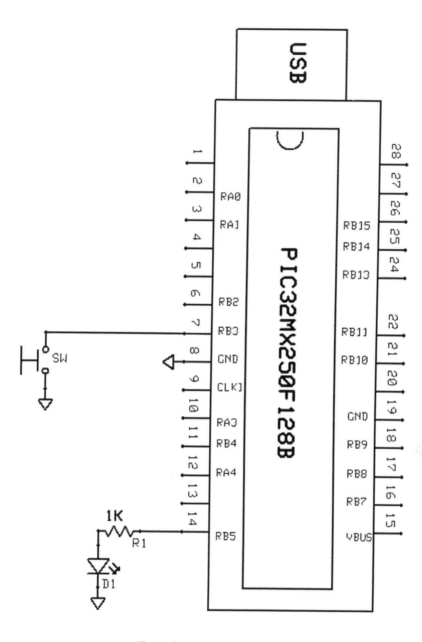

Figure 3-5: Experiment 1 Schematic

A full schematic for Experiment 1 is shown (Figure 3-5). Make sure that the ground potential for your prototype circuit is the same ground reference used by

Beginner's Guide to Programming the PIC32

the Microstick II, so connect the Solderless breadboard ground to pin 8 of Microstick II.

Experiment 1- The Software

Ok, the hardware is in place, lets discuss the software. First bring up MPLAB X, then open project and Navigate to Chapter 3 code folder, open exercise 1 subfolder and then double click the MPLAB X project icon exercise 1 pushbutton. At this point you may have several projects opened in MPLAB X, make sure this one is the main project. It should be in bold compared to others.

The application uses a single C source file (main.c), and a single header file System.h as was done in Chapter 2. Both of which can be examined in editor by double clicking on the title in project view. The main function references System.h (#include "System.h"). This header will determine the specific PIC32MX processor in use as well as configuration fuse setting. These settings are: internal CPU clock selection of 8Mhz, no external oscillator, no clock switch over, no watchdog operation, and finally use of RA3 as a digital/analog pin versus an external crystal pin input.

In the main function the application executes I/O initialization and then proceeds to enter a continuous loop testing whether pin 7 RB3 is high or low. If high the processor turns off the LED connected to RB5, if low it turns on the LED. A flow diagram of this application is shown in Figure 3-6.

A new define statement is introduced to label the switch SW as PORTBbits.RB3. Several new SFR settings are also introduced. One is TRISbits.TRISB3 =1 to set RB3 as an input. The other is CNPUBbits.CNPUB3 =1 set the corresponding weak pull up for RB3 input.

The main continuous loop uses an IF ELSE decision, testing the switch or SW (PORTBbit.RB3 value) condition and then setting or resetting the LED. This is done without any delays. They are just not necessary here as we want to respond as immediately as possible to the switch closure and report it via a LED.

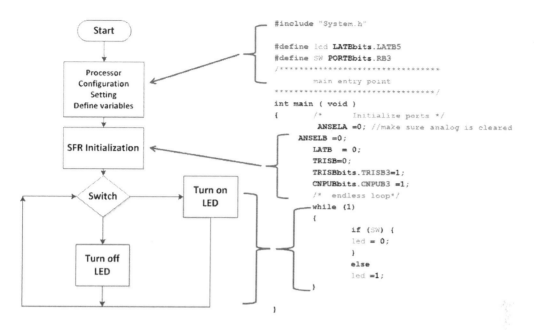

Figure 3-6: Experiment1 Software operations

Experiment 1 Execution

Let's go ahead and begin debug. At this point we have a fairly good idea of what will happen during execution given that the prototype hardware is configured correctly, and the source code and processor configuration are in good shape. The PIC32MX will turn on the LED connected to pin 14 RB5 of the Microstick II while the pushbutton is depressed. Simulation will not help much here as we are using an actual hardware input.

Examine dashboard and make sure that Starter Kit is selected as debugger, also make sure your Microstick II is plugged into the USB of your PC. (See Figure 3-7).

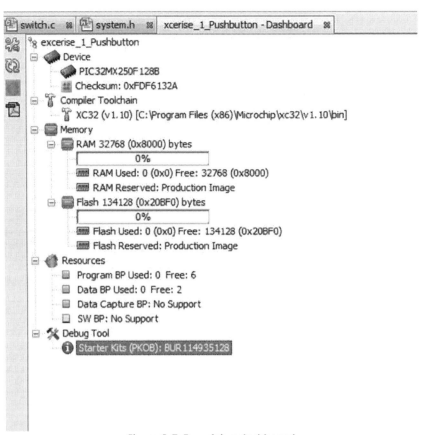

Figure 3-7: Examining dashboard

Launch the application by clicking on the Debug Project icon shown in Figure 3-8. This will put the PICkit 3 in communication mode and send data back from the running PIC32 to the MPLABX screen.

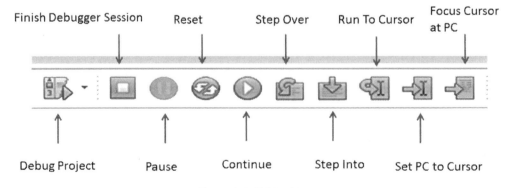

Figure 3-8: IDE Tool Bar

You should see "Starter Kit on Board" tab in the Output display. Figure 3-9 shows that tab.

Figure 3-9: Start kit activated

There should also be a debugger console as shown in Figure 3-10. Note the debugger progress bar is showing system activity running in lower right hand corner of IDE (not shown here).

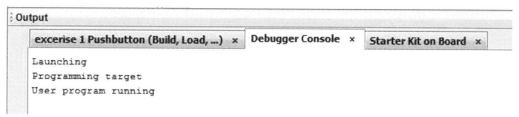

Figure 3-10: Debugger Console Running

Let's set a break point and use the watch window to make sure the debugger is reading the switch correctly. Pause the Debugger by clicking on the Pause icon. Set the Watch Window to display PORTB. You should see the contents of Figure 3-11.

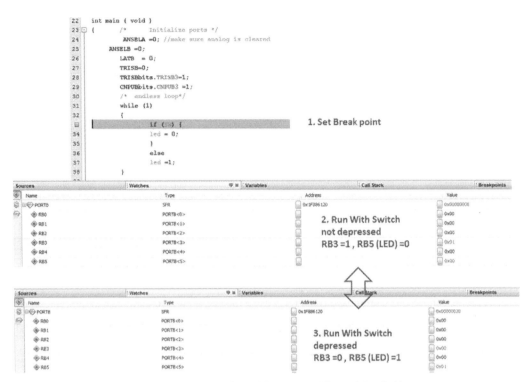

Figure 3-11: debugging with watch set for RB3 (switch)

Ok, if your watch captures the results of Figure 3-11 your hardware is working. If not go back and review the schematic and assembly shown earlier, you may have a hardware wiring error. Assuming all is correct simply toggle the breakpoint setting off using the toolbar and run. Every time you depress the pushbutton the LED should light.

Experiment 2 -The Hardware

Let's move on to Experiment 2. Here's where we must now account for switch bounce to develop an accurate count of switches closures. We will build a five stage LED counter. This counter will count manual pushbutton depressions. With a 5 stage counter we should count 31 push button depressions before overflow. This should be good enough for this experiment. Again the push button is connected to RB3. The LEDs are connected through current limiting 1K resistors to RB7, RB8, RB9, RB10, and RB11. A schematic is provided.

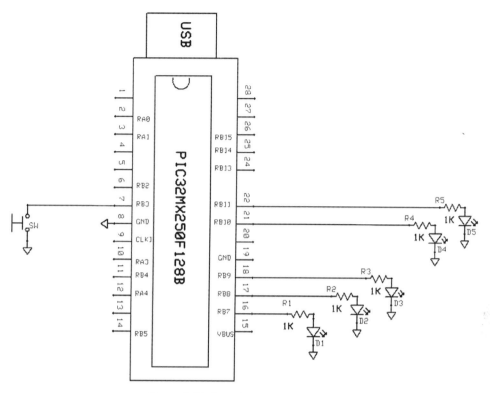

Figure 3-12: Experiment 2 Schematic

Experiment 2 –How it works

One of the simplest ways to debounce is to sample the switch until the signal is stable and no more bounces are detected. How long to delay for debounce requires some investigation. However, several hundred milliseconds is usually plenty long enough, while still reacting fast enough that the user won't notice it.

Our solution is to sample the line at a 100 millisecond rate, test for first switch depression, and then retest for stable depression after a fixed delay (using delay with a debounce counter). If the depression is stable after a given amount of counts then a switch depression is accepted and the "debounce" is achieved.

Experiment 2 -The Software

Ok, the hardware is in place, let's discuss the software. Open MPLAB X and navigate to Chapter 3 code folder open exercise #2 pushbutton folders, and then double click the MPLAB project icon exercise 2 pushbutton. At this point you may have several projects opened in MPLAB X, make sure this one is the main project. It should be in bold compared to others.

The application again uses a single C source file (main.c) and single reference file System.h , as in Exercise 1, it is a main function that references generic PIC32MX headers/peripherals use (#include "System.h"). This header will determine the specific PIC32MX processor in use and configuration settings. These settings are principally the internal CPU clock selection of 8Mhz, no external oscillator, no clock switch over, no watchdog operation, and finally use of RA3 as a digital/analog pin versus an external crystal pin input.

A define statement is used to label the switch SW as PORTBbits.RB3. This is the input pin connected to the pushbutton. Other defines are associated with limits used in the delay function. Again we use the delay function and it will be instrumental in debouncing the pushbutton.

In the main function the application executes I/O initialization. We initially set all pins from analog to digital, ADSELA and ANSELB to all zeros. We clear LATB and initially set all of PORTB pins to digital out. Remember that RB7, RB8, RB9, RB10, RB11 are tied to LEDs and will represent our binary counter. We then set TRISbits.TRISB3 =1 to enable RB3 as an input. The other final setting is CNPUBbits.CNPUB3 =1 to set the corresponding weak pull up for RB3 input.

The main code then proceeds to enter a continuous loop testing whether pin 7 RB3 (SW) is low representing a pushbutton depression. If low we delay and increment a debounce counter. If the SW is still low after 3 counts we interpret this as a final stable pushbutton depression and then perform the binary count. This binary count is then output to LATB to drive the LEDs. A flow diagram of this application is shown in Figure 3-13.

The main continuous loop uses three nested decision constructs. The first IF statement tests the SW condition, the second nested IF ELSE tests the debounce count and then determines to increment the binary count or not. The third nested IF does a check on the incremented count and resets the count if it exceeds a maximum count of 4096. Note there is also a view count variable. The actual counter "binarycount" increments by 0x7f (remember our LSB is RB7).

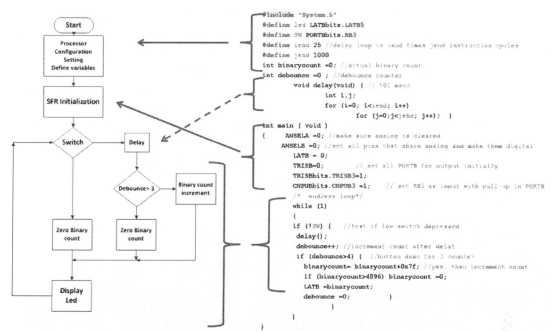

Figure 3-13: Experiment 2 Software

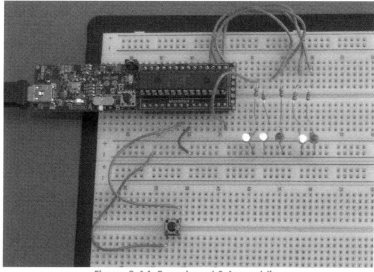

Figure 3-14: Experiment 2 Assemblies

Experiment 2 Execution

At this point we have a fairly good idea of what will happen during execution given that the prototype hardware is configured correctly, and the hardware wiring is correct. Remove any previous breakpoints, reset the processor, and hit run. The PIC32MX should increment the five stage binary counter with every depression of the pushbutton connected to pin 7 RB3 of the Microstick II. Each depression and release of the pushbutton should result in only one increment of the binary counter. Any problems at this point must be attributed to hardware so double check with the provided schematic and assembly photos.

Review of Key Application problem with using Switches:

- Mechanical switches play an important and extensive role in practically every computer, microprocessor and microcontroller application.

- Mechanical switches are inexpensive, simple and reliable. However, switches can be very noisy electrically.

- The apparent noise is caused by the closing and opening action that seldom results in a clean electrical transition. The connection makes and breaks several, perhaps even hundreds, of times before the final switch state settles.

- The problem is known as switch bounce. Some of the intermittent activity is due to the switch contacts actually bouncing off each other. Imagine slapping two billiard balls together. The hard non-resilient material doesn't absorb the kinetic energy of motion.

Review of Source Code Key Features:

- We introduced a technique to debounce a mechanical switch through successively sampling of the switch input over a period of time.
- A new define statement is introduced to label the switch SW as PORTBbits.RB3
- Several new SFR settings are also introduced.
    - One is TRISbits.TRISB3 =1 to set RB3 as an input.

- The other CNPUBbits.CNPUB3 =1 to set the corresponding weak pull up for RB3 input.

Review of MPLAB Debug Features:

- Watch –used to check SFR PORTB value to ensure the processor is correctly reading the input switch

Exercise:

1. Try decreasing the debounce count in Experiment 2 to a smaller value. Note that that as the value decreases switch debounce is not smooth and which each depression and release we incur a number of counts.

2. Set a WATCH to view count and check the values of the binary count are correct for the number of pushbutton activations that occurred.

3. Try to add a second button to RB2 and use this button to decrement the running count.

4. Try using the Microchip peripheral library to set the CNPUB SFR.

Chapter 4 - PIC32 Interrupts and Change Detection

This chapter is purposely positioned after Chapter 3 to introduce PIC32 interrupt capability. As you may recall in Chapter 3 we focused on exercises that sensed an input switch to activate either a single LED or increment a binary LED display. In all those chapter exercises we used an simple approach to have the microcontroller test the switch input to determine if the switch was depressed or not. This technique is also known as "polling". Polling is a time consuming activity for the microcontroller, in that, no matter what else it may have to do as part of its application, it has to revisit the switch input often enough so as not to miss switch activation. Clearly this is an inefficient use of the microcontroller processing power, and gets even more labor some as additional switches are involved. There has to be a better way. The way to resolve this is the use of interrupts.

We will introduce the use of interrupts with the PIC32, by revisiting the previous chapter exercise requirements and then addressing them with interrupts. The PIC32 interrupt capability that can best handle the problem are external interrupts (INT0) and the internal change notice detection port B interrupt (CNB). The exercises will cover both. To reinforce the interrupt concept our new exercises will demonstrate successful switch closure detection using interrupts, while at the same time that the microcontroller is busy performing an additional task blinking an LED. The use of interrupts occurs throughout the book as we introduce new peripheral capabilities.

The PIC32 interrupt can be an external signal to the microcontroller or an internal generated microcontroller peripheral signal. Both these interrupt sources indicate an event that needs immediate PIC32 attention. The PIC32 responds to this high-priority condition by immediately interrupting current code execution, saving current processing state, and then executing a small program called an interrupt handler (interrupt service routine, ISR) to deal with the condition. The interruption is temporary, and after the interrupt handler finishes, the PIC32 resumes execution of the code just prior to the interrupt.

It is up to the user to decide when to use interrupts in their microcontroller application. Using interrupts increases the response time of the microcontroller to achieve more "real-time" operation, in that, the internal peripheral or outside event immediately alerts the microcontroller when service is needed. By deciding not to use interrupts the user may run the risk of missing a response to critical outside events. For example the loss of communications data, or an air-bag trip delay, or anti-lock breaking and speed control delays. One of the distinct advantages of PIC Microcontrollers is guaranteed latency time between interrupt occurrence and detection by the microcontroller. This makes the PIC a very desirable microcontroller to use in real time applications.

The PIC32MX250F128B Interrupt Capability

The PIC32MX250F128B Interrupt controller generates interrupt requests in response to interrupt events from external sources and peripheral modules. The interrupt control module exists externally to the PIC32 CPU logic and prioritizes the interrupt events before presenting them to the CPU.

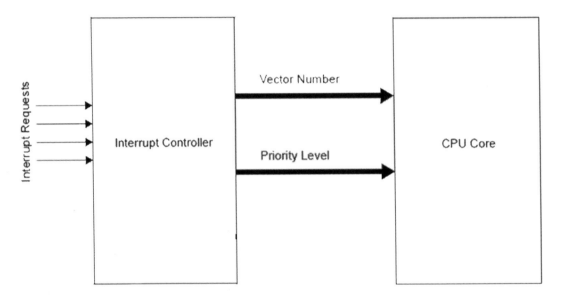

Figure 4-1: Interrupt Logic Block Diagram

The PIC32MX has an interrupt controller with two modes of operation:
- Multi-vectored
    - Interrupt flags(s) are associated with a vector
    - Each flag is set upon an interrupts condition and must be individually cleared before another similar interrupt can occur
    - Up to 44 unique vectors (each assigned a unique interrupt source), includes the internal peripherals and the 5 external interrupts
- Single Vectored
    - Used primarily with an overriding Operating System
    - All interrupts become associated with a single vector

- Within the interrupt service software has to resolve the interrupt source

We will focus on the Multi-vector interrupt mode in this book. This mode is closely connected with Microchip peripheral capability, and the XC32 C compiler has a template that exists for each vector, it makes the interrupt design straightforward.

However in this mode several other system capabilities are realized:
- Seven user-selectable priority levels for each vectors
- Four user-selectable sub priority levels within each priority
- A dedicated shadow set for all priority levels

Let's explain these features in turn in detail.

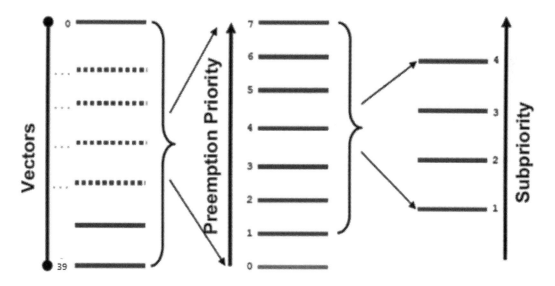

Figure 4-2: Interrupt Priority Scheme

Interrupt vectors are assigned priority. This makes sense in that multiple interrupts can be present in the design and may occur during operation simultaneously. By prioritizing each one will determine when each interrupt will receive attention from the PIC32 in order of importance. These priorities are assigned as part of the interrupt service template. Zero represents the highest priority and seven the lowest. The PIC32 interrupt controller also allows for sub-priorities (1 to 4) within the existing higher priority system (0-39) if two interrupts of the same higher priority "fire off" at the same time. Given the limited number of priorities needed within this book we will ignore sub-priorities settings. The dedicated shadow register option is uniquely limited to high speed interrupts, where the internal CPU registers can be masked to a single alternate set for faster interrupt service versus saving the register in memory. Again, this is a unique special feature and will not be covered in this text.

The PIC32 interrupt ISR

An interrupt ISR is needed to handle the match event interrupt generated by either the external interrupt or the internal generated interrupt. The simplest method to define an interrupt handler is to use an X32 'C' compiler macro and requires only the vector symbol corresponding to the interrupt source (see symbols below) and an interrupt priority level, IPLx. Note, that the the name of the ISR can be whatever you want to name it. Here's the overall syntax:

Void __ISR (<vector number symbol>, <priority>) HandlerName (void)

Here's an example applied to internal peripheral Timer 5 with a priority of 2.

Void __ISR (_TIMER_5_VECTOR, IPL2) myHandler (void)

As mentioned there are vector symbols for each of the 44 unique vectors. Some of these vectors are illustrated here.

Vector Number Symbol Definitions

```
_CORE_TIMER_VECTOR              _INPUT_CAPTURE_5_VECTOR
_CORE_SOFTWARE_0_VECTOR         _OUTPUT_COMPARE_5_VECTOR
_CORE_SOFTWARE_1_VECTOR         _SPI1_VECTOR
_EXTERNAL_0_VECTOR              _UART1_VECTOR
_TIMER_1_VECTOR                 _I2C1_VECTOR
_INPUT_CAPTURE_1_VECTOR         _CHANGE_NOTICE_VECTOR
_OUTPUT_COMPARE_1_VECTOR        _ADC_VECTOR
_EXTERNAL_1_VECTOR              _PMP_VECTOR
_TIMER_2_VECTOR                 _COMPARATOR_1_VECTOR
_INPUT_CAPTURE_2_VECTOR         _COMPARATOR_2_VECTOR
_OUTPUT_COMPARE_2_VECTOR        _SPI2_VECTOR
_EXTERNAL_2_VECTOR              _UART2_VECTOR
_TIMER_3_VECTOR                 _I2C2_VECTOR
_INPUT_CAPTURE_3_VECTOR         _FAIL_SAFE_MONITOR_VECTOR
_OUTPUT_COMPARE_3_VECTOR        _RTCC_VECTOR
_EXTERNAL_3_VECTOR              _DMA0_VECTOR
_TIMER_4_VECTOR                 _DMA1_VECTOR
_INPUT_CAPTURE_4_VECTOR         _DMA2_VECTOR
_OUTPUT_COMPARE_4_VECTOR        _DMA3_VECTOR
_EXTERNAL_4_VECTOR              _FCE_VECTOR
_TIMER_5_VECTOR
```

Figure 4-3: List of ISR Vector Symbols

Designing for interrupt services

Within the interrupt service an interrupt flag is not automatically cleared by the hardware. You must clear the interrupt flag manually as part of the routine. Failure to do so will cause the interrupt to be locked out for any future occurrences. You must also set priority for the interrupt and finally enable the interrupt. Sub-priority is optional. The priority you set as part of the ISR definition must match the priority you use when you set up the peripheral. Here's the generic format, because of the extensive interrupt capabilities the PIC32MX has 2 Interrupt Flag 32 bit SFRs (IFS0, IFS1), 2 Interrupt Enable 32 bit SFRs (IEC0, IEC1) and 8 Priority Setting 32 bit SFRs (IPC0, IPC1, IPC2, IPC3, IPC4, IPC5, IPC6, IPC7). Refer to PIC32MX250F128B datasheet interrupt register map to identify specific IFS, IPC, and IEC SFRs for your peripheral of interest. We will examine more specific cases in the exercises to follow:

- Example Clear interrupt flag: IFSXbits.Flag_Name = 0. X =0,1
- Example Set priority IPCXBits.Priority_Settings=0-7 (must match ISR priority setting). X =0-7
- Example Enable Interrupt IECXbits.Perpherial_ Enable bit =1. X =0,1

You can save any registers/variables you access in the ISR's code manually. Variables modified by an interrupt should be tagged with the volatile 'C' keyword. Avoid use of calling any functions during ISR as this will increase the interrupt latency time. For interrupts to work you must set the interrupt mode, using the following PIC32 library call, as follows:

INTEnableSystemMultiVectoredInt ();

Let's get started with the exercises to clarify some of these concepts.

Exercise 1 INT0 Interrupt

Wire the prototype as follows: RB5 (pin 14) LED 1 is driven by interrupt based upon switch closure to RB7 (pin 16) (INT0). RB8 (pin 17) LED 2 is driven on/off during non-interrupt operation. All LEDs have a series 1K resistor to ground. Assembled prototypes are similar to those shown in Chapter 3.

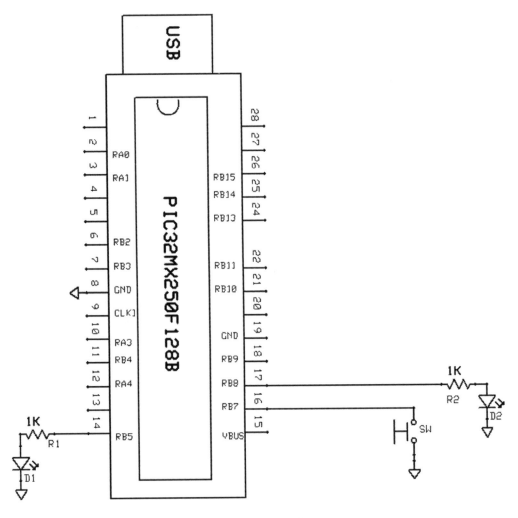

Figure 4-4: Schematic for Exercise 1

The external interrupt INT0 (pin 16 RB7) is used with an internal pull resistor as the input switch (using SFR setting CNPUBbits.CNPUB7=1). As a normal course of operation the PIC32 is toggling RB8 pin 17, that also has a led connected through a series resistor, once a second. The PIC32MX has five external interrupts (INT0, INT1, INT2, INT3, INT4), each of which can interrupt based upon a rising or trailing pulse edge to the signal. INT0 is easier to use and already device programmed for pin 16. We will use a trailing edge interrupt, trailing edge is needed since switch input is normally high because of internal pull up resistor.

Beginner's Guide to Programming the PIC32

Set the interrupt mode as follows:

1. INTEnableSystemMultiVectoredInt ();

The INT0 set edge polarity to falling using SFR INTCON as follows:

2. INTCONbits.INT0EP =0; //falling edge

Then the follows needs to occur:

   //Clears interrupt flag. SFR IFS0 contains the appropriate bit
3. IFS0bits.INT0IF =0;   //clear interrupt flag

Set interrupt priority SFR IPC0

4. IPC0bits.INT0IP =2;   //set interrupt priority to 2

Enable interrupt   SFR IEC0

5. IEC0bits.INT0IE =1;   //enable interrupts

Finally configuring the interrupt service routine (matching priority setting with earlier declaration) as:

6. Void __ISR (_EXTERNAL_0_VECTOR, ipl2) INT0_Handler (void)

Write the ISR code using the following library functions. Make sure to clear the interrupt flag to allow interrupt to reoccur.

7. INTClearFlag (INT_SOURCE_EX_INT (0)); // clear the interrupt flag
8. PORTToggleBits (IOPORT_B, led1); // toggle the LED

The following Figure 4-5 shows the entire code segment and outlines all the critical sections.

```c
#include "System.h"
#define led1 LATBbits.LATB5
#define led2 LATBbits.LATB8
#define SW  PORTBbits.RB7    //INT0 pin 16
#define iend 25  //delay loop is iend times jend instruction cycles
#define jend 1000
void __ISR(_EXTERNAL_0_VECTOR , ipl2) INT0_Handler(void)
{
    // clear the interrupt flag
    INTClearFlag(INT_SOURCE_EX_INT(0));
    PORTToggleBits(IOPORT_B, led1);
}
    void delay(void) { // 501 msec
        int i,j;
        for (i=0; i<iend; i++)
            for (j=0;j<jend; j++); }
int main ( void )
{   ANSELA =0; //make sure analog is cleared
    ANSELB =0; //set all pins that share analog and make them digital
    LATB = 0;
    TRISB=0;          // set all PORTB for output initially
    TRISBbits.TRISB7=1;  // enable for switch input
    CNPUBbits.CNPUB7=1; //change notice pull-up
    // enable multi-vector interrupts
    INTEnableSystemMultiVectoredInt();
// set INT0 edge polarity
    INTCONbits.INT0EP =0;  //falling edge
// clears interrupt flag, setsinterrupt priority, enablesinterrupt
    IFS0bits.INT0IF =0;  //clear interrupt flag
    IPC0bits.INT0IP =2;  //set interrupt priority to 2
    IEC0bits.INT0IE =1;   //enable interrupts
    //~~~~~~~~~~~~~~~~~~~~~~~~~~~~~~~~~~~~~~~~~~~~~~~~~~
    while (1) {
        PORTToggleBits(IOPORT_B, led2);
        delay();
    }
}
```

- System constant definitions
- Interrupt Service Routine
- SFR configurations
1. Non-interrupt main code

Figure 4-5: Exercise 1 INT0 code

Build the prototype as per the schematic. Open MPLABX and navigate to folder excerise3.5. Open project exercise 1. Connect the Microstick II, and then build the debug project. The project should run. Witness the LED 2 blinking. Press button and witness LED 1 toggle. Any issues could be the result of wiring or Microchip II placement on the prototype board. Check and redo.

Exercise 2 Change Notice (CN) Interrupt

Wire the prototype as follows. Assembled prototypes are similar to those shown in Chapter 3. We will move the switch back to the original position used in Chapter 3, RB3 (pin 7), since INT0 is not going to be needed. LED 1 (RB5 pin 14) is driven by a change notice detection interrupt associated with switch closure to RB3 (pin 7). RB8 (pin 17) LED 2 is driven on/off during non-interrupt operation. All LEDs have a series 1K resistor to ground.

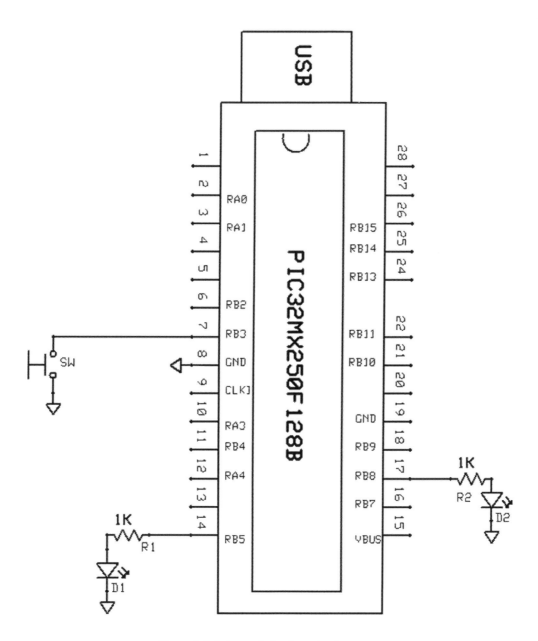

Figure 4-6: Schematic for Exercise 2

The Change Notice is enabled for RB3 pin 7 (using SFR setting CNENBbits.CNIEB3 =1, change notice enable). The pin is also configured with an internal pull resistor for the input switch (using SFR setting CNPUBbits.CNPUB3=1). Normal switch reading is a high level and when pressed shows a low level. Without the pull up the switch input would float and the reading would be indeterminate. Under Change Notice the PIC32 peripheral logic monitors assigned pins for change in state. This can be used as a source of interrupt. Once the state of RB3 changes, it can be recognized as a switch closure detection.

In this exercise, as the main course of operation the PIC32 is toggling RB8 pin 17 every .5 seconds. RB8 has an LED connected through a series resistor to ground. Once change detection occurs on RB3 we clear the interrupt flag associated with Change Detection, clear the change detection logic by reading the port and then toggle RB5 pin 14.

RB5 also has a LED connected through a series resistor to ground. Setting up the interrupt process is similar to the exercise shown earlier, with the only changes for Change Detection configuration. Let's review the key steps:

Set the PIC32MX for multi-vector interrupt using the following library function:

1. INTEnableSystemMultiVectoredInt ();

Configure RB3 as an input with internal pull up resistor using SFRTRISB and CNPUB as follows:

2. TRISBbits.TRISB3=1;   // enable for switch input
3. CNPUBbits.CNPU3 =1; // use internal pull up with RB3

Configure and enable Change Detection Logic on RB3 using SFR CNEN, and CNCON.

4. CNENBbits.CNIEB3 =1; //change notice configure for RB3
5. CNCONBbits.ON =1; //change notice control

Read PORTB to clear mismatch on change notice pins.

6. Temp = mPORTBRead();

Clear Change Detection interrupt flag. SFR IFS1 contains the appropriate bit.

7. IFS1bits.CNBIF=0; //clear interrupt flag

Set Change Detection interrupt priority to 2 using SFR IPC8.

8. IPC8bits.CNIP =2; //set interrupt priority to 2

Enable Change Detection interrupt enable using SFR IEC1.

9. IEC1bits.CNBIE=1;   //enable interrupts

Finally configure the interrupt service routine (matching priority setting with earlier declaration).

10. Void __ISR (_CHANGE_NOTICE_VECTOR, ipl2) ChangeNotice_Handler (void)

Write the ISR code. Make sure to clear the interrupt flag to allow interrupt to reoccur.

11. mCNBClearIntFlag();

The ISR code reads the current detection status and port value and compares them to determine they are both one (indicating a button release). It then clears the Change Detection logic with a read and toggles LED1.

```
If ((CNSTATBbits.CNSTATB3 ==1) && (PORTBbits.RB3==1)) {
  PORTToggleBits (IOPORT_B, led1);
  // clear the mismatch condition
   Temp = mPORTBRead ();
}
```

The following figure shows the entire code segment and outlines all the critical sections

```
void __ISR(_CHANGE_NOTICE_VECTOR, ipl2) ChangeNotice_Handler(void)
{
    // clear the interrupt flag
    mCNBClearIntFlag();
    if ((CNSTATBbits.CNSTATB3 ==1) && (PORTBbits.RB3==1)) {
        PORTToggleBits(IOPORT_B, led1);
        // clear the mismatch condition
        temp = mPORTBRead();
    }
}
int main ( void )
{
    ANSELA =0; //make sure analog is cleared
    ANSELB =0; //set all pins that share analog and make them digital
    LATB = 0;
    TRISB=0;         // set all PORTB for output initially
    TRISBbits.TRISB3=1;   // enable for switch input
    CNPUBbits.CNPUB3=1; //change notice pull-up
    CNENBbits.CNIEB3 =1;  //change notice enable
    CNCONBbits.ON =1; //change notice control
    //~~~~~~~~~~~~~~~~~~~~~~~~~~~~~~~~~~~~~~~~~~~~~~~~~~~~~~~~
    // Read port(s) to clear mismatch on change notice pins
    temp = mPORTBRead();
    //~~~~~~~~~~~~~~~~~~~~~~~~~~~~~~~~~~~~~~~~~~~~~~~~~~~~~~~~
    // enable multi-vector interrupts
    INTEnableSystemMultiVectoredInt();
    //1. clears CN interrupt flag
    //2. sets CN interrupt priority
    //3. enables CN interrupt
    IFS1bits.CNBIF=0;  //clear interrupt flag
    IPC8bits.CNIP =2;  //set interrupt priority to 2
    IEC1bits.CNBIE=1;  //enable interrupts
    //~~~~~~~~~~~~~~~~~~~~~~~~~~~~~~~~~~~~~~~~~~~~~~~~~~~~~~~~
    while (1){   //main task
        delay();
        PORTToggleBits(IOPORT_B, led2);
    }
}
```

- Interrupt Service Routine (top block)
- SFR and Change Detection setup and Configurations Enable interrupts (middle block)
- Non-interrupt main code (bottom block)

Figure 4-7: Exercise 2 Code

Build the prototype as per the schematic. Open MPLABX and navigate to folder excerise3.5. Open project exercise 2. Connect Microstick II, and then build the debug project. The project should run. Witness the LED 2 blinking. Depress button and witness LED 1 toggle. Any issues could be the result of wiring or Microchip II placement on the prototype board. Check and redo

Review of PIC32 Interrupts:

- The PIC32 interrupt can be an external signal (INT0, 1, 2, 3, 4) to the microcontroller or an internal generated microcontroller peripheral signal (40 total).

- Interrupts provide a processing alternative for PIC32 in effectively dealing with real time events. The PIC32 is designed for low latency in recognizing and launching interrupt service requests (ISR)

- The PIC32 responds to interrupts by interrupting current code execution, saving current processing state, and then executing a small program called an interrupt handler (interrupt service routine, ISR) to deal with the condition. The interruption is temporary, and after the interrupt handler finishes, the PIC32 resumes execution of the code just prior to the interrupt.

- The PIC32 has two modes of interrupt operation Single and Multi-vector

- Under Multi-Vector there are 40 sources of interrupts, each of which is designated its own assigned interrupt vector.

- There are seven levels of priority (7- Highest, 0 lowest) that are assignable to an interrupt, and allow the PIC32 to arbitrate which interrupt to process in the event multiple interrupts occur simultaneously.

- Within each priority level there are 4 levels of sub priority to enable further arbitration.

- The PIC32 supports a "fast context" interrupt feature where all core registers are shadowed automatically. This allows the ISR to immediately execute without separate variable storage and retrieval. Because there is only one set of shadow registers only one interrupt can use this capability in an application.

- You can save any registers/variables you access in the ISR's code manually. Variables modified by an interrupt should be tagged with the volatile 'C' keyword. Avoid use of calling any functions during ISR as this will increase the interrupt latency time

Review of Source Code Key Features:

- Within the interrupt service an interrupt flag is not automatically cleared by the hardware. You must clear the interrupt flag manually as part of the routine. Failure to do so will cause the interrupt to be locked out for any future occurrences.
- You must also set priority for the interrupt and finally enable the interrupt. Sub-priority is optional. The priority you set as part of the ISR definition must match the priority you use when you set up the peripheral.
- The PIC32MX has 2 Interrupt Flag 32 bit SFR (IFS0, IFS1), 2 Interrupt Enable 32 bit SFRs (IEC0, IEC1) and 8 Priority Setting 32 bit SFRs (IPC0, IPC1, IPC2, IPC3, IPC4, IPC5, IPC6, IPC7). Refer to PIC32MX250F128B datasheet interrupt register map to identify specific IFS, IPC, and IEC SFR for your peripheral of interest.

- Common Interrupt SFR configurations:
    - Example Clear interrupt flag: IFSXbits.Flag_Name = 0. X =0,1
    - Example Set priority IPCXBits.Priority_Settings=0-7 (must match ISR priority setting). X =0-7
    - Example Enable Interrupt IECXbits.Perpherial_ Enable bit =1. X =0,1
- For interrupts to work you must set the interrupt mode, using the following PIC32 library call for multi-vectored interrupts, as follows:

    INTEnableSystemMultiVectoredInt ();

Exercise:

1. Try redoing Exercise 2 with a different switch input rather the RB3
2. Combine INT0 and Change Detection interrupts for a two interrupt example. Introduce a second switch for INT0. Use different priorities for either. Have both interrupts toggle LED 2.
3. Leveraging (2) introduce a LED 3 for just INT0 to toggle. Leave LED 2 for Change Detection.

Chapter 5 – Using the ADC

In this chapter we will explore the use of the PIC32MX Analog to Digital Converter or ADC. The ADC is an important peripheral for the PIC32MX. It allows capture and measurement of external analog values into the PIC32MX, and then formats these values to digital numbers for PIC32MX processing. The ADC peripheral enhances the operations of the microcontroller in sensing and responding to real world conditions. The outside digital values, for example, could be a temperature reading originating from a temperature sensor, pressure from an industrial MEMS sensor, or light conditions sensed from a photocell. As you may imagine this capability allows the microcontroller to address a very large number of real world applications: thermometer, barometric, or blood pressure and lighting control to name a few.

In this chapter we will explore in detail how the ADC works, what the PIC32MX offers in terms of an ADC, and then how to set up and use the features of the PIC32MX ADC.
In our experiments we will use the ADC to perform a simple voltage measurement and then examine and interpret its digital results. For an analog input voltage source we will use an adjustable potentiometer.

We will also introduce a set of generic ADC library 'C' functions that are easily understandable, and can be reused in other experiments. These library functions enable ADC initialization, retrieve a reading from the ADC, provide average successive readings from the ADC for better accuracy, and finally allow for formatting the ADC binary output to ASCII (the American Standard Code for Information Interchange) to support text output .

However, in order to use the ADC we need to understand it's operational functionally, associated constraints, and how to set it up correctly. A quick review of basic Analog to Digital Operations is in order (see Figure 5-1).

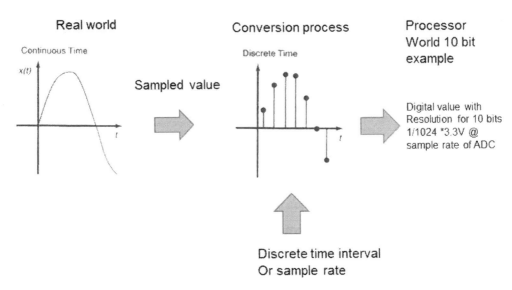

Figure 5-1: Analog to Digital Process

In Figure 5-1 the analog environment (typical of an input sensor value) is shown as a continuous voltage output over time. As environmental conditions change the sensor output value also changes smoothly without interruption. The microcontroller, however, is a digital device; it operates only with digital numbers, and can only make a measurement periodically (as part of a normal processing loop). From the microcontroller viewpoint the continuous sensor signal is now a discrete set of values sampled at a fixed rate. The values are approximate values to a true analog value given the simple fact that the digital value number size itself is limited in resolution, or number of bits. The PIC32MX ADC can generate a 10 bit value.

A 10 bit resolution provides a 1/1024 *3.3VDC voltage resolution or 3 millivolts (remember that the PIC32MX is a +3.3VDC part). The PIC32MX ADC resolution accuracy is good to +-1 LSB of the digital word size. The accuracy can be improved by averaging successive reading (assuming input does not change during averaging) to better than an LSB. A rule of thumb with this technology is to insure that the signal (by dynamically adjusting its amplitude) gets as large as possible to the ADC without exceeding full scale to achieve the best signal resolution during sample measurement. In our experiments we will not control any signal amplitude and only measure the raw output directly.

Let's now discuss the fixed measurement rate or sample rate. As mentioned earlier, by using a fixed sample rate you are measuring the signal periodically. With periodical sampling the risk is that the signal may change between the periodic samples significantly and the microcontroller will not "see" or therefore measure and respond to these changes. Another rule of thumb in engineering is that the sample rate should be high enough to capture the highest frequency of the signal. To be more precise the rule states that the sample rate = 2X Highest Input frequency (also known as Shannon's Rule). In most cases (outside of audio) the sensor outputs are typically very low in frequency content so that a high sample rate isn't necessary. However, the highest sample rate for the PIC32MX with 10 bit ADC is 1.1 MSPS (million samples per second).

Let's cover some more basics a before moving on with the experiments. Figure 5-2 shows some of the steps that are involved with an ADC operation, beyond the sampling rate and digital resolution. The ADC operation is broken into two discrete steps: acquisition time (sampling and holding the signal) and then conversion (performing the ADC process on the sampled and held signal). In order to collect and hold an analog signal for conversion a sample and hold capacitor is used. This capacitor is internal to the PIC32MX electronics. The capacitor follows the analog input signal and then upon a sample command stores a voltage capture of the analog value on the capacitor for the conversion process. This is denoted as the sample and hold process.

While the signal is captured in the capacitor an analog to digital conversion process takes place. The conversion uses 10 clock cycles (equivalent to resulting output word size), to perform the conversion. This type of conversion process is known as successive approximation, where an internal comparator is used to measure the difference between the analog sampled input and an internal generated ramp. For each clock (one of the 10) a digital decision is made based upon comparison of sample against the ramp, and a digital bit value is generated and then stored as part of the final digital word result.

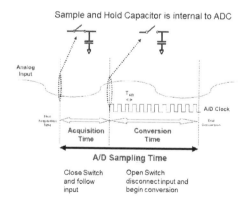

Figure 5-2: ADC Sampling Time

Beginner's Guide to Programming the PIC32

The PIC32 10-bit Analog-to-Digital Converter (ADC) includes the following features:
- Successive Approximation Register (SAR) conversion
- Up to 16 analog input pins
- External voltage reference input pins
- One Sample-and-Hold Amplifier (SHA)
- Automatic Channel Scan mode
- Selectable conversion trigger source
- 16-word conversion result buffer
- Selectable Buffer Fill modes
- Eight conversion result format options
- Operation during CPU Sleep and Idle modes

Figure 5-3 illustrates a block diagram of the 10-bit ADC. The 10-bit ADC can have up to 16 analog input pins, AN0 through AN15. The actual number of analog input pins and external voltage reference input configuration will depend on the specific PIC32 device. Our specific PIC32MX, pre-configured in hardware with the Microstick II, doesn't have 16 analog channels but is limited to 10. These ten channels are AN0, AN1, AN3, AN4, AN5, AN6, AN9, AN10, AN11, and AN12.

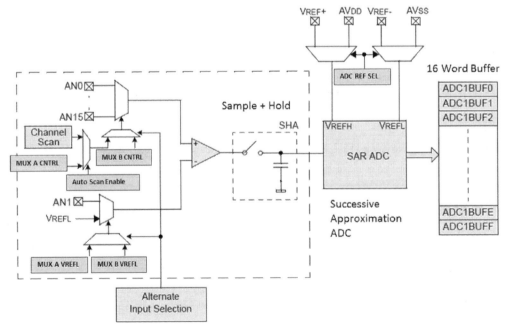

Figure 5-3: 10 bit PIC32MX ADC Configuration

There are two options for the ADC external voltage reference connection that are selectable through ADC reference selection (ADC REF SEL). In our case we use the power +3.3VDC and GND for reference.

The analog inputs can be essentially connected to two functional multiplexers A and B (MUX A and MUX B) for direct channel selection. The analog input multiplexers can be switched between two sets of analog inputs between conversions using MUX A or MUX B Channel Control. We only use MUX A. The low reference offset into the Sample and Hold amplifier is also selectable through MUX A or B VERFL. The VERFL is set to ground to eliminate any offset for conversions during in our exercise.

The Analog Input Scan mode can be used to sequentially convert user-specified channels. Channel Select specifies which analog input channels will be included in the scanning sequence. Channel scan is not enabled for our exercise, since we are using only one channel conversion.

The 10-bit ADC SAR (Successive Approximation Output) is connected to a 16-word result buffer. Each 10-bit result can be converted to one of eight selectable 32-bit output formats when it is read from the result buffer. For our exercise we use unsigned integer magnitude 16 bit.

All these settings are accomplished in the ADC module by configuring in software a total of 6 SFR Control and Status registers. Let's review this SFR content. In summary we want to be able to set the ADC to single channel mode operation, no scanning, integer output, and manual mode trigger.

Using the SFR setting for integer unsigned format for 10 bits, the signal digital output using this format varies from 0x0000 to 0x03FF. This is typical for most single voltage input sensor values where 0 represents 0.0 volts and 0X03FF represents +3.3 volts.

The AD1CON1, AD2CON2 and AD3CON3 SFR registers used to control the operation of the ADC module are listed below.
AD1CON1 SFR bits

- ADON =1 turns on the ADC
- FORM bits
    - 8 selectable data format
    - 000 sets 16 bit Integer

- (ADC Sample Auto-Start) ASAM bit =1 allows sampling to begin when SAMP is set to 1
- (ADC Sample Enable Bit) SAMP bit =1 forces sampling on the designated input and then auto-convert when SSRC =111
- DONE bit
    - 0 = ADC conversion is complete
    - 1 = ADC conversion is not complete or not started

AD1CON2 SFR bits

- VCFG bits =000 sets ADC +REF to +3.3VDC and the –REF to 0.0 VDC
- (Input Scan Select) CSCNA bit =0 no scanning
- (Buffer Fill Status Bit) BUFS bit = 0 always start filling output buffer at 0 position
- BUFM ADC result Buffer Mode Select
- (Alternate Input Mode Select) ALTS bit =0 uses S/H 0

AD1CON3 SFR bits

- ADRC bit =0 sets conversion clock to internal CPU clock
- SAMC bits = 11111 sets largest sample time
- ADCS bits = 00111111 sets largest conversion time

AD1CHS SFR bits –sets input channel 0 (S/H 0) analog channels select

- CH0SA Channel A =000 positive input or AN0
- CH0NA Channel A =0 negative input of –REF
- CH0SB Channel B =000 –not used
- CH0SB Channel B =0 negative input of –REF -not used

AD1CSSL SFR bits –input scan select Low

- CSS = 0x0001 for AN0

Hardware Description

The hardware schematic is given in Figure 5-4. Here we use an analog 10K potentiometer to function as a settable input analog source for analog input channel 0 or AN0 pin 2 of PIC32MX250F128B. The 10K potentiometer has one terminal tied to an external +3.3 VDC voltage source.

The Microstick II has two bare pads GND and VDD (or on schematic TP1). Your options are to connect Microstick power to the breadboard via the TP1 pad or furnish a separate supply voltage for this experiment. An important note in using TP1 is that there is a limitation in how much current this connection can supply. Remember that the microcontroller itself is also using this connection. Its needs will vary based upon oscillator setting and output drive current. As a rule of thumb as your prototype experiments grow beyond the 100ma mark you may want to use an external supply for your power source.

In this exercise, since we are only using a potentiometer, we will connect to the VDD (TP1) for the +3.3VDC of the potentiometer and also connect pin 8 of Microstick II ground to the ground for the potentiometer. The wiper of the potentiometer is tied to the AN0 input pin 2. You must remove jumper J3 to disconnect the User LED from this pin. With the potentiometer we can vary the input voltage to AN0 to any value between +3.3VDC to ground for our ADC experiment.

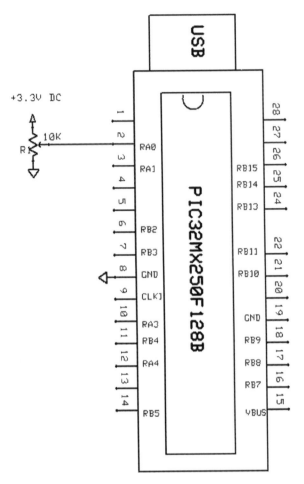

Figure 5-4: Schematic

Software Description

In this chapter we will again use multiple source code files for our project. In order to use these multiple source code files or ".c" files the corresponding ".h" files or header files also need to be included in the project, and referenced in main. The function of the ".h" file is to reference the C functions, variables, defines, or macros as legitimate C entities that can be used in the main function.

The advantages of this approach are an extension of main functionality and modularizing of any required additional C functionality for use by the end application, without the need to integrate the entire C code into main. This approach also provides convenient reuse of code for other applications that may need similar functionality.

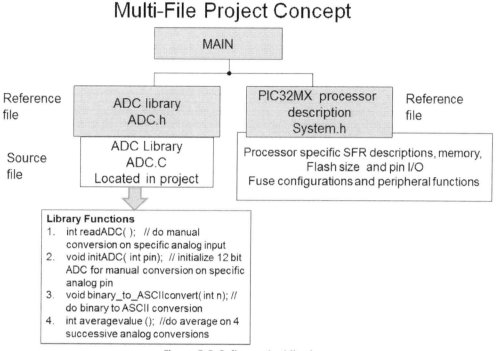

Figure 5-5: Software Architecture

We mentioned the need and purpose of System.h in Chapter 2. Let's move to the new elements. Specifically we need a separate source library file for ADC processing. These functions will need to support ADC peripheral initialization, ADC conversion on demand, ADC output conversion to ASCII, and support a moving average process that works on successively ADC outputs to increase resolution. Again, all of these functions are important and have significant potential value for reuse in other applications.

As mentioned earlier, in order for main to incorporate other software modules, a "# include" directive must be used. This "# include" will be used to reference an ADC.h file. For every ".h" file there must exist a ".c" file that contains the actual source code. From this perspective our software project organization looks like the following.

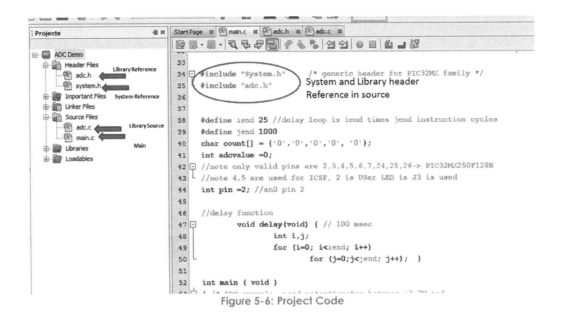

Figure 5-6: Project Code

The actual library functions for ADC are as follows

- **Void initADC (int pin)** –this function initializes the ADC for 12 bit manual conversion using the "ANpin" as the single input pin. Pin possible settings are AN0, AN1, AN2, AN3, AN4, AN5, AN6, AN9, AN10, AN11, and AN12. Note AN4 and AN5 have dual purpose as debugger controls for ICSP. Only one channel can be initialized at a time.

- **Int readADC ()** –this functions forces a manual ADC conversion on the designated ANpin.

- **Void binary_to_ASCIIconvert (int n)** - this functions converts the n binary value to its ASCII equivalent. The ASCII equivalent resides in BCD variables listed below

- **Int averagevalue ()** – this function is a high level function that automatically does four ADC conversions successfully in a row and averages the total.

In addition there are five variable locations where the library holds the output from a binary to ASCII conversion

- char bcd10000 -ASCII data value for 10,000 position

- char bcd1000 -ASCII data value for 1000 position

- char bcd100   -ASCII data value for 100 position

- char bcdtens - ASCII data value for 10 position

- char bcdunits – ASCII data value for units position

The complete main software flow and code is shown in Figure 5-7.

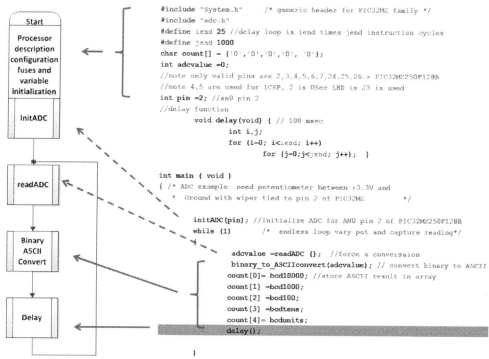

Figure 5-7: Main Code

Running the Experiment

Plug the Microstick II into solderless breadboard. Add the potentiometer to circuit as shown in schematic. Connect one end of the potentiometer to ground and the other to +3.3VDC external power supply. Make sure ground is also connected to pin 8 of Microstick II. Connect Microstick II to PC USB port.

Figure 5-8: ADC Prototype

Navigate to the Exercise 1 folder for Chapter 5. Open this folder and double click on the Microchip project icon. MPLABX should open and the output window should show that the Microstick II is connected. Build and program the Microstick II under debug.

Open a watch window and set watch for two variables adcvalue and count array. Note you can expand the count array by clicking on the + sign associated with the variable.

Set a break point at delay function in main, and run from reset condition.

For a quick check the PIC32MX should stop at delay function, and the watch variables should be populated. Adjust the pot and run the program again, the variable should change in value. If the variable is not changing check your wiring against the schematic. There is an error in your wiring.

Let's conduct some final verification. Compare your adcvalue against count array values. You can use a hexadecimal to decimal conversion on adcvalue using a calculator, or right click on adcvalue in watch window and set properties to decimal and make a direct comparison.

Another and even more complete form of verification is to incorporate the use of a voltmeter and connect this voltmeter between potentiometer wiper and grounds for the voltage measurement.

Take the decimal count from either count array or adcvalue, multiply using a calculator, by the LSB value (+3.3V/1024) and see if your ADC count matches the voltmeter by an LSB.

Congratulations you have an ADC capability and a reusable library! We will be using this library in subsequent chapters.

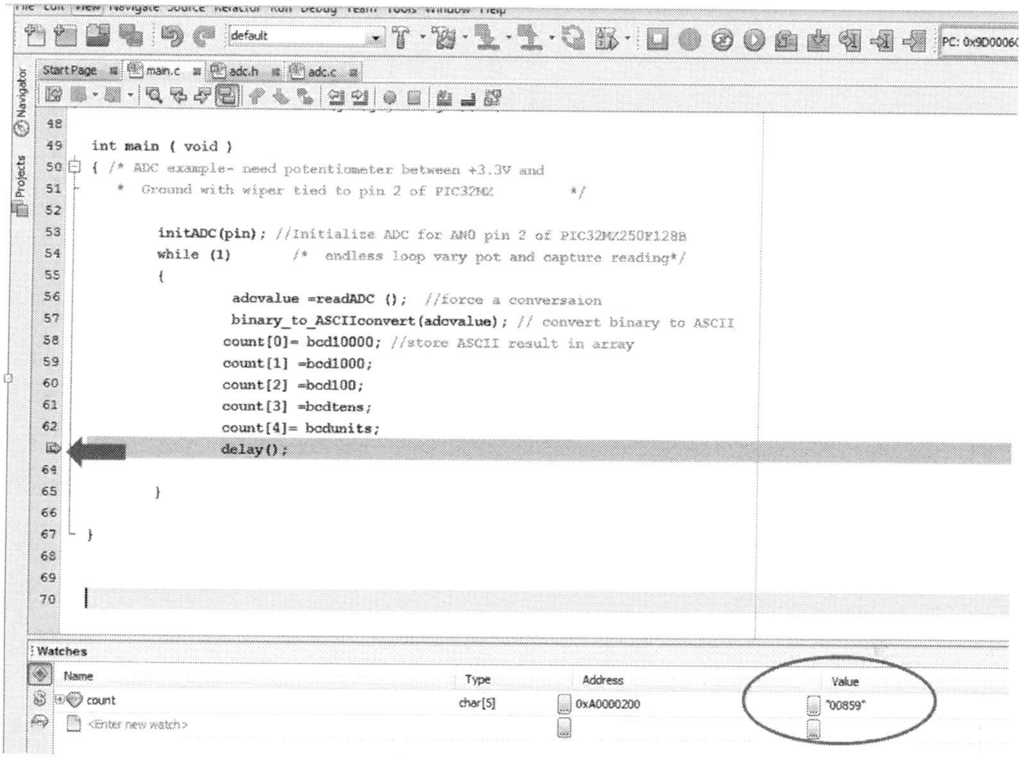

Figure 5-9: ADC Demo

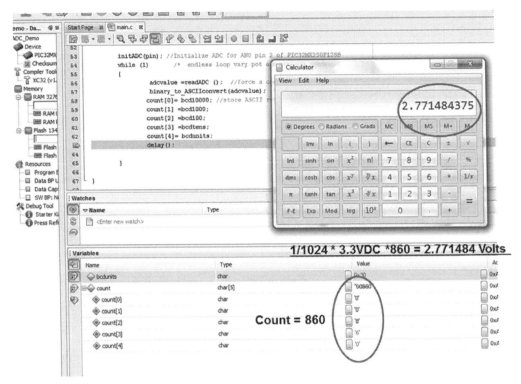

Figure 5-10: Calculating Voltage using Watch and Calculator

Figure 5-11: Validating ADC conversion with actual reading

Review of Key Application using ADC:

- PIC32MX ADC is configurable for 10 bit operation.

- Multiple S/H are required for some multi-channel simultaneous data collection

- 10 bit can run up to 1.1 MSPS

- There are a number of SFR settings to configure the ADC

    - AD1CON1, AD2CON2 and AD3CON3 SFR registers control the operation of the ADC module.

    - AD1CHS –input analog channel select

    - AD1CSSL –input channel scanning

- The AVDD and ADVSS are tied the Microstick II to +3.3VDC and ground. A 10 bit resolution essentially provides a 1/1024 *3.3VDC voltage resolution or 3 millivolts (reminder that the PIC32MX is a +3.3VDC part).

Review of Source Code Key Features:

- Use of include directive makes the code highly modular and it this case an ADC driver library (ADC.h) was introduced.
- Project contains multiple files for each driver where the explicit driver function are identified the body ADC .c
- Main just includes the .h its needs and can directly access all driver code
- Global variable identified in ADC library is accessible by all code. In this case it is the BCD variables.

Exercise :

1. Try changing to different analog port. Make sure to move the pot wiper to the new analog input

2. Try using the averaging function in the library. Compare its output results to using the normal conversion routine and also compare using a voltmeter to read the wiper to ground voltage. Which one is more accurate? Remember that to determine voltage multiply the ADC output count by 3 millivolts.

Chapter 6 – Using an LCD Display with the Microstick II

In this chapter we will use an external LCD (Liquid Crystal Display) with the Microstick II. There is both software (in the form of a library) and hardware interfaces that need to be considered. An LCD really enhances the operations of a microcontroller by displaying status, providing important application feedback and facilitating operational control with the user.

The LCD we will use supports a multiple line text display of 3 row by 16 characters for a total of 48 characters. The LCD has its own integrated built-in controller that performs a lot of the "heavy lifting" for the more mundane aspects of display operation. For instance the LCD controller drives the individual display segments on the display; automatically generates the display character segments for an ASCII text input, and finally, responds to a fixed command and data interface to an external microcontroller (which in our case will be the Microstick II).

In this chapter we will incorporate an LCD connection scheme and software library that can be easily reusable for your future projects. To solidify our understanding we will conduct several experiments using this configuration. The first experiment will be concerned with just setting up and operating the LCD for the first time. In the second experiment, we will use the LCD in conjunction with the ADC library introduced in Chapter 4 to function as a simple voltmeter; outputting and displaying ADC data in a voltage format. Let's get started!

Hardware Description

The LCD is a +3.3VDC powered 16x3 character LCD Display that is manufactured by Electronic Assembly Corporation. It is very similar in control to other standard LCDs but has an extended feature of built in contrast control (eliminating the need for any external potentiometers). It also consumes very low power (typically 250 micro amps @ +3.3VDC). The LCD is designed for compact hand-held devices; it is also extremely compact, ultra-flat, and also supports a "DIP-like" pin out, making it ideal for breadboard prototyping.

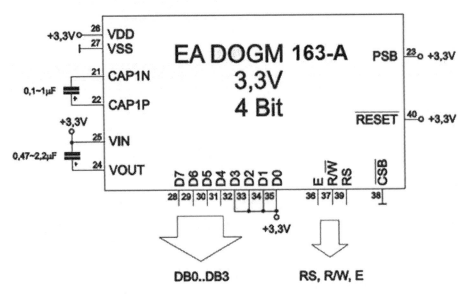

Figure 6-1: EADOGM LCD Display

The LCD display interface is a 4-bit data and 2- bit command digital interface. The 4 bit interface is optional to the normal 8 bit interface, and requires two 4 bit nibble writes rather than one byte write to the LCD. This interface is selected to simplify prototype wiring and pin requirements with the Microstick II. Another simplification is to not have the Microstick II read from the LCD but only to write to it. To facilitate this, the LCD library forces the Microstick II PIC32MX to delay between its command/data transactions with the LCD controller with enough time to allow the LCD time to respond without having to read and check its busy status.

A schematic that illustrates the Microstick II connection to the LCD is shown in Figure 6-2.

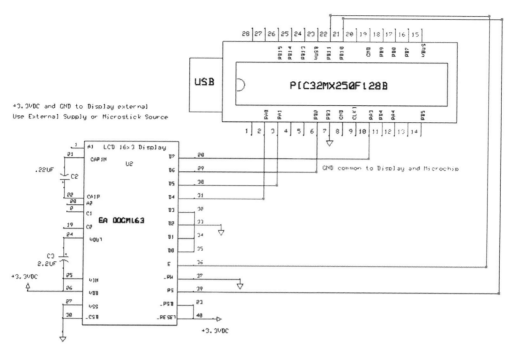

Figure 6-2: LCD and Microstick II Schematic

Note that a separate +3.3VDC supply is required to drive the LCD since the Microstick II itself does not supply external power. Another option for +3.3VDC is to connect to the TP1 via on the Microstick II, since the LCD has a very small current need. The ground connection between the LCD and Microchip must be common and connected to pin 8 as shown.

The LCD does require use of separate electronic components for its operation. Both a 2.2uf and .22uf capacitors (C3, C2) are needed (see Figure 6-2) for contrast control and internal voltage regulation on the LCD. Note that the 2.2uf is polarized –so follow those marking assuring the positive side is connected to pin 25 and 26.

The LCD pin arrangement, as indicated earlier, is configured as a 40 pin dip (with most of the bottom row pins missing). As you look at the display from its top, the bottom left hand side pin is pin 1, and the bottom right is pin 20. The top row of pins, from left to right are pins 40 to 21. Please follow along with all the connections shown in the schematic and do a visual comparison with the picture of the prototype. Also note that the Electronic Assembly LCD display comes packaged with a protective film on the display side. This should be removed (peeled off) for un-obstructive viewing.

In summary, the LCD 4 bit parallel data interface to Microstick II is as follows:

- LCD D7 to Microstick II pin RA3
- LCD D6 to Microstick II pin RB2
- LCD D5 to Microstick II pin RA1
- LCD D4 to Microstick II pin RA0

The command interface is facilitated with two command lines

- LCD command/data control RS to Microstick II pin RB10
- LCD strobe in E to Microstick II pin RB11

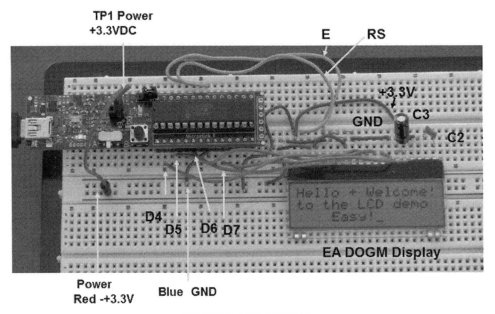

Figure 6-3: LCD prototype

Software Library Description

So much for hardware, now here's the fun part. Let's learn the software functions for driving the LCD and how to use them. Here are the five we will be working with. Their titles pretty much capture what they do. Remember the LCD only accepts ASCII characters for data. The LCD, as mentioned earlier, is organized as a display with 0-47 (48) character positions or similarly a 16x3 display matrix. Again, this refers to 16 columns by 3 rows. The first row contains positions 0-15, the second 16-31, the third 32 -47. The library allows complete freedom to position a character anywhere in the display matrix. It also auto-increments the character position when you are writing one then one character (like a string) at a time.

- LCD_Initialize () - initializes the LCD display and makes it ready for output. This needs to be called before any other LCD function is called. Keep in mind initialization is a two-step process, first the PIC32MX must initialize it ports for output connection to the LCD and then once this is achieved it then executes the LCD initialization command sequence. Also keep in mind that the library is written specifically to support the connections illustrated in the schematic. Any deviation from this wiring will cause the LCD library not to work correctly.

The LCD requires specific initialization sequence for its proper operation. There is also a strict command set for PIC32MX to interact with the LCD for writing command and data and also positioning the LCD cursor. These functions are performed by using the supplied 'C' library functions. Both an initialization sequence and command interface using elements of the 'C' library are shown for illustration. You are encouraged to review this code (in project under Graphics.c) and compare it against the LCD display data sheet that comes with the display (see Figure 6-4).

Initialization Definitions in Graphics Library (Graphics.c)

```
//LCD Hardware interface
#define LCDPORT LATA
#define E    LATBbits.LATB11
#define RS   LATBbits.LATB10
//EA DOG Initialization Instructions 4 bit 3.3V
#define FUNCT1    0x29   //DL-4bits 2 lines Table select =1
#define BIAS      0x15   // BIAS 1:5 3 line LCD
#define PWRCNTRL  0x55   //booster on  constrast c5 ,set c4
#define FOLLOW    0x6D   //set booster follower and gain
#define CONST     0x78   // set constrast c3,c2,c1
#define FUNCT2    0x28   //switch back to table =0
#define DISPLAY   0x0F   //display on/off --display on, cursor on, cursor blink
#define CLEAR     0x01   //delete display --cursor at home
#define ENTRY     0x06   //cursor auto-increment
```

Initialization Sequence in Graphics Library (Graphics.c)

```
// FUNCT1       0x29   //DL-4bits 3 lines Table select =1
LCDWrite(FUNCT1,0);
// BIAS         0x14   // BIAS 1:5 2 line LCD
LCDWrite(BIAS,0);
// PWRCNTRL 0x55       //booster on  constrast c5 ,set c4
LCDWrite(PWRCNTRL,0);
// FOLLOW       0x6D   //set booster follower and gain
LCDWrite(FOLLOW,0);
// CONST        0x78   // set constrast c3,c2,c1
LCDWrite(CONST,0);
// FUNCT2       0x28   //switch back to table =0
LCDWrite(FUNCT2,0);
// DISPLAY      0x0F //display on/off --display on, cursor on, cursor blink
LCDWrite(DISPLAY,0);
// CLEAR        0x01   //delete display --cursor at home
LCDWrite(CLEAR,0);
// ENTRY        0x06   //cursor auto-increment
LCDWrite(ENTRY,0);
```

Figure 6-4: LCD Code for Initialization

LCD library Functions Continued

- **clear_display ()** - clears display and places cursor at row 1 column 1 (top left corner)
- **position_cursor (position)** - places cursor at designated position, a number from 0-47
- **write_string_LCD (string name)** - writes a string to display at current cursor position
- **write_character_LCD (character)** - writes a single character to display at current cursor position

LCD Library Support Functions

Delays – software delay is an important component for Library operation both for initialization and for writing data and commands. To accomplish this, the library uses delay.h and a delay.c files. The delay library uses Timer 1 and adjusts the timer period countdown using the current CPU clock setting. The good news here is the delay self-adjusts based upon CPU clock. The only exception is the 1us delay. The period here is too short for Timer1 and a software implementation was required. This function will need to be adjusted as the clock changes. The delay functions are shown here. The operations are self-explanatory, but again refer to the display data sheet to better understand why these delays are needed.

- **void delay_1us(void)** - 1 microsecond delay used in LCD for E
- **void Delay_10us(t)** - t =3 for 30 microsecond delay used between commands
- **void Delay_msec(t)** - t =5 millisecond used for Busy delay
- **void Delay_msec(t)** - t =40 millisecond used for power up delay

Let's see what the project view looks like and where these library functions reside and how they are referenced. We covered System.h in earlier chapters. Let's now review LCD and Delay. The library files reside as source code C files and as header file ".H" in the project. The header files are referenced within Main as shown.

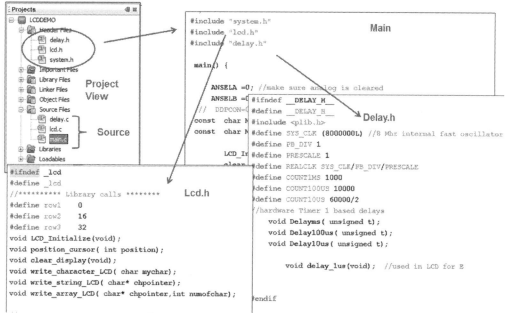

Figure 6-5: Library Configuration in Project

Experiment #1 – LCDDEMO

An example project LCDDEMO is supplied. This project shows how to use each of the above LCD functions and then how to call them from your MAIN file. Let's walk through the flowchart and then the code. LCDDEMO starts by declaring two strings: message# 1 and message# 2. Each message is configured to be fewer than 16 characters to keep the display nice and neat, that is, keeping each message to within a single display row.

The next step is to initialize the LCD and clear display contents before use. The next step after this is to position the cursor to beginning of row one (cursor position =0) and write out message #1. We then position the cursor to the beginning of row two (cursor position =16) and write out message #2. Just to keep things interesting we then position the cursor to the middle of line three (cursor position = 36) and write out five characters.

The LCD is initialized is so that the cursor itself is not visible, and the cursor position automatically advances to the next position when data is written to the display. Any data for display that is written to the LCD must be in ASCII (American Standard for Information Interchange) format.

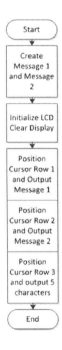

Figure 6-6: LCDDEMO Flowchart

Running LCDDEMO

Navigate to Chapter 6 Exercise 1 folder. Double click on the LCDDEMO microchip project demo. Make sure your Microstick II and LCD are wired correctly and an external +3.3VDC display is available for the LCD and that the LCD and Microstick II both share a common ground. Hook up the Microstick II via USB, build the project, and then download the code. A copy of LCDDEMO main code is shown (Figure 6-7) along with the resulting Experimenter LCD display, once that code is executed.

Notice that we have both **#include "lcd.h",** and **#include "Delay.h"** at the top of the file to allow us to reference both the LCD library and delay library. Try to match up the flowchart with the code in main to understand the library syntax. To help you along, there are lots of comments in the code. If you have not successfully reach this point, with the display functioning, please review your wiring by following the provided schematic point to point, along with the photo of the prototype. Also make sure you understand the pin arrangements of the LCD.

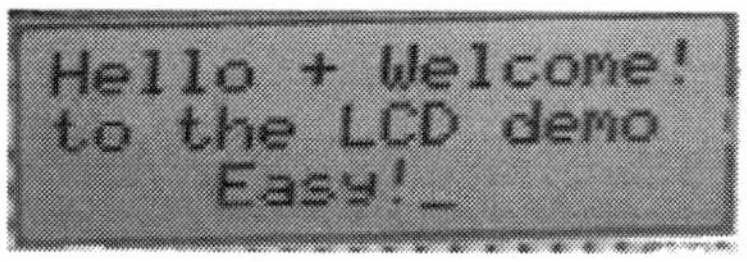

```
main() {

    ANSELA =0; //make sure analog is cleared
    ANSELB =0;
//   DDPCON=0x00;
const  char Message1[] ="Hello + Welcome!";
const  char Message2[] ="to the LCD demo";

    LCD_Initialize();                //initialize
    clear_display();                 //clear the LCD display

    position_cursor(0);              //position cursor to row 1
    write_string_LCD((char*)Message1);   //outputs Message1 to display

    position_cursor(16);             //position cursor to row 2
    write_string_LCD((char*)Message2);   //outputs Message 2 to display

    position_cursor(36);             //position cursor to position 36
    write_character_LCD('E');
    write_character_LCD('a');
    write_character_LCD('s');        //outputs "Easy!" using single characters
    write_character_LCD('y');
    write_character_LCD('!');

    while(1);    //permanent wait

//end of code        // set IO as outputs for PORTB

}              LCD Main Demo
```

Figure 6-7: LCD Demo

Writing to the Experimenter LCD using these library functions is straightforward, and to prove this to yourself try modifying the code to configure your own display messages---you will find that writing your own messages to the LCD is a snap!

Experiment #2 –A Simple Voltmeter

This experiment will expand our Microstick II application allowing us to build a simple voltmeter that uses both the LCD and ADC libraries we learned so far. The PIC24 Analog to Digital Converter (ADC) can be configured as a 12 bit converter that has up to 11 input channels and performs conversions up to 500K per sec. The conversion automatically converts a voltage signal at the input of an ADC pin between 0 volts to +3.3 Volts to a numeric twelve bit value 0 to 4095 representing that voltage. This capability allows the PIC32MX to measure external voltage sources (i.e. like sensor outputs). Let's try it.

We use the library to configure an I/O pin (in this case pin 26 or AN9) as an ADC input before performing a conversion. The wiring schematic is shown. The only difference between this schematic and the one shown earlier is the addition of a 10K potentiometer.

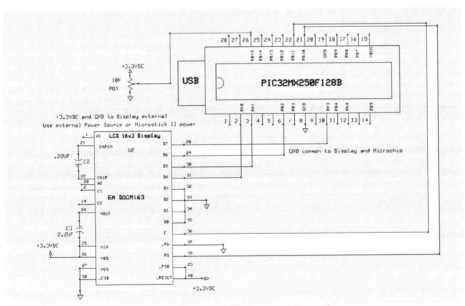

Figure 6-8: Voltmeter schematic

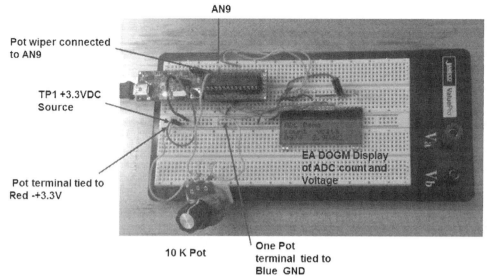
Figure 6-9: voltmeter prototype

The ADC library functions used are as the same as the ones discussed in Chapter 4:

- **Void initADC (int pin)** –this function initializes the ADC for 12 bit manual conversion using the "pin" as the single input pin. Pin possible settings are AN0, AN1, AN2, AN3, AN4, AN5, AN6, AN9, AN10, AN11, and AN12. Note AN4 and AN5 have dual purpose as debugger controls for ICSP. Only one channel can be initialized at a time.

- **Int readADC ()** –this functions forces a manual ADC conversion on the designated pin.

- **Void binary_to_ASCIIconvert (int n)** - this functions converts the n binary value to its ASCII equivalent. The ASCII equivalent resides in BCD global variables

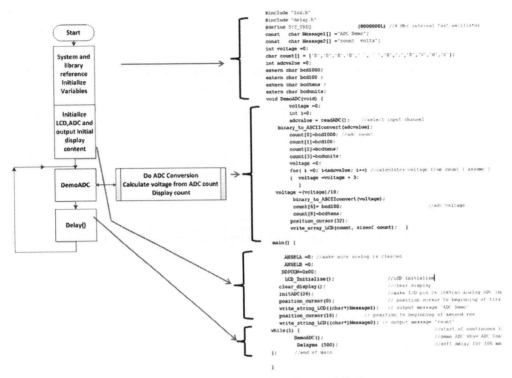

Figure 6-10: Voltmeter Flowchart and Code

Running the Voltmeter

Keep the LCD wired in place from Experiment 1. Modify the prototype with the potentiometer as show in the schematic. Connect one end of the potentiometer to ground and the other to +3.3VDC external power supply. Make sure ground is also connected to pin 8 of Microstick II. Connect Microstick II to PC USB port. Plug Microstick II into Solderless breadboard. Navigate to the exercise folder for Chapter 6. Open this folder and then open folder for voltmeter. Double click on the PIC32HVOLTMETERDemo project. MPLAB lab should be opened.

Build and program the Microstick II under debug, and the output window should show that the Microstick II is connected, click on RUN. You should get a display similar to Figure 6-9 with the ADC count and voltage associated with that count simultaneously displayed. Try changing the pot value to change readings ---full CCW of pot should be zero count and zero volts, full CW of pot should be 4096 count and +3.3V. If readings are erratic you have a wiring error with pot or are on the wrong analog input pin.

Congratulations you have a simple ADC voltmeter capability.

Review of Key Application using LCD:
- LCD is a configurable intelligent device that is commanded by the PIC24H during application.

Review of Source Code Key Features:
- Use of include directive makes the code highly modular and it this case an ADC driver library was introduced.
- Project contains multiple files for each driver where the explicit driver function are identified in .h and the body in .c
- Main just includes the .h its needs and can directly access all driver code
- Global variable identified before main function is accessible by all code
- Here we introduced another library function the LCD.

Exercise:
1. Try changing first experiment's messages to your own and rerun the experiment.

2. Determine where BCD1000 reside in source. How is it accessible by main?

3. Make experiment 2 use the pot wiper to a different analog input

4. Try using the averaging function from the ADC library in experiment 2. Note that display readings are more stable with averaging.

Chapter 7 – Using Timers and Timer Interrupts

In this chapter we will describe and experiment with the timer capabilities inherent in the PIC32MX250F128B. We will also introduce the important concept of an interrupt as used with the timer.

The timer is a hardware counter that expedites counting and delay operations that would normally be time consuming and inefficient if done in software. The timer peripheral is a very significant PIC32MX peripheral capability. If you recall from earlier chapters we used a software based delay function, by simply looping in software to exact a given delay. With the PIC32MX timers we can measure external count/external timing events as direct counts or internal PIC32MX CPU clock counts. All this can occur in hardware with minimal need for software intervention. Once a timer is set up it performs its functions with little to no oversight by the CPU, and then only alerts the CPU once a count match event has occurred--truly a very powerful and important capability.
The PIC32MX device family has two different types of timers, depending on the device variant. With certain exceptions, all of the timers have the same functional circuitry. The timers are broadly classified into two types, namely:

Type A: (Timer 1)
- Asynchronous timer/counter with a built-in oscillator
- Operational during CPU Sleep mode
- Software selectable prescalers 1:1, 1:8, 1:64 and 1:256

Type B: (Timers 2, 3, 4, 5)
- Ability to form a 32-bit timer/counter
- Software prescalers 1:1, 1:2, 1:4, 1:8, 1:16, 1:32, 1:64 and 1:256
- Event trigger capability (Timer 3 only ADC event trigger)

All timer modules include the following common features:
- 16-bit timer/counter
- Software-selectable internal or external clock source
- Programmable interrupt generation and priority
- Gated external pulse counter

Timer rates and initial count settings are all configurable under software. We will examine the use of B type timers in a series of simple LED blinking operations to understand how to set up both 16 and 32 bit timer operations, verify any count operations initially through calculation, then simulate and finally real time debug. We will also (as mentioned earlier) introduce the concept of interrupt here as is applies to timers operation.

The 16 bit Timer Peripheral

A high level block diagram of the B Type 16 bit timer is shown in Figure 7-1. This block diagram simplifies the overall peripheral representation, and in fact, eliminates a number of other important features that the peripheral contains, i.e. sync, gating, but for our purposes this level of representation is adequate.

Remember there are a total of five of these peripherals in the PIC32MX. Operations for all these timers are similar.

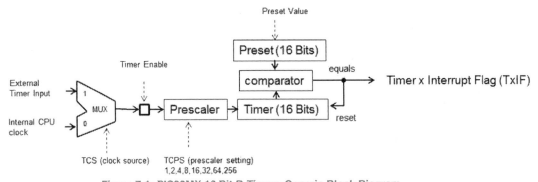

Figure 7-1: PIC32MX 16 Bit B Timers Generic Block Diagram

Each element within the timer has its own set of SFR based controls that the microcontroller needs to initialize. Starting on the left side of the block diagram is the input clock source selection for the timer controlled by TCS (timer clock select bit). The TCS setting allows for either external clock or the internal PIC32MX peripheral clock. The PIC32MX internal clock with our current fuse setting is 8 MHz—which is the clock we will use for our demo.

The next control is Timer Enable that turns on the timer. The next stage is the prescaler that allows the clock to be scaled down by factors of 1, 8, 16, 32, 64, and 256. The prescaler is set using control bits TCPS (Timer Clock Prescaler set). All the control bits reside in an internal control register designated Timer Control (T1CON for Timer 1). The final element within the block diagram is the preset register and the timer register both of which are 16 bits wide for total counts of 65,536.

The microcontroller writes to the preset register to set the upper count range for the timer. During timer operation the preset value is constantly compared against the running count value of the timer for a match condition. If a match occurs a Timer Interrupt flag is latched to alert the microcontroller. Another action that occurs as a result of this latch is the reset of timer register--- This automatically starts the next timing cycle.

A Timer Interrupt flag can also cause an interrupt to PIC32MX if the timer interrupt has been enabled. We will cover the Interrupts in more detail in the Timer 2 Experiment, but will touch on it here briefly to understand its use with timers.

The interrupt is a construct that allows the CPU to immediately process a specific function call, no matter where the CPU is operating in its current code space. The interrupt does this with a minimal delay, and also with minimal disruption to overall code operation. This is the most efficient way for the microcontroller to perform periodic processing without any delays. In a sense the timer alerts the CPU that a timer timeout has occurred so that the CPU can perform some critical time-dependent activity without delay. It is a way for the system to respond to and work towards real time operations.

Interrupt processing works as follows, you must have a source of interrupt (in our case the timer), you must enable the interrupt operation with the PIC32MX interrupt controller, and finally you must have a specific function call, which is designated the interrupt service routine, that will be invoked and executed once the interrupt occurs. There are some more details about specifics of the service routine, its construction, and what it minimally must do. We will cover this shortly.

Once a timer interrupt condition occurs, the current state (program counter, status registers, and flags) of the CPU is saved or "pushed" onto a stack (first in first out ram buffer internal to the PIC32MX). The timer interrupt service routine is then invoked for CPU execution to handle the timer interrupt condition. Once the CPU exits the service routine, the CPU state prior to the interrupt is restored or "popped" from the stack. The CPU then resumes its normal operations. It in a sense interrupts to allow for an "on demand" peripheral processing service.

Hardware Description

We have seen this specific hardware setup in Chapter 2, here we will use it again. It is the familiar blinking LED prototype, but in this experiment the software is entirely new. We will be using entire new techniques with the timer. The same software I/O setup will be common in all the experiments. In these experiments we want to connect a single LED to bit 5 of PORTB or RB5 and set RB5 to be a digital output.

RB5 is pin number 14 on the PIC32MX250F128B. The device is a 28 pin DIP, so there are 14 pins on each side of the chip. Looking down on the chip there is a notch in the package to indicate pin 1, all the pins are numbered 1 to 28 and go counter clockwise on the device. This would make pin 14 or RB5 the last pin on the bottom right hand side of the chip.

A schematic is shown in Figure 7-2. The PIC32MX digital output can drive up to 25 milliamps on a digital output. We don't need all this current to turn on a LED (3 milliamps is more than enough), so we use a current limiting resistor (1K) in series with the LED. The circuit is basically a series circuit where RB5 tied to one side of the resistor, the other side of the resistor is connected to LED anode, and finally the LED cathode is tied to ground. In this way when we make RB5 =1 the LED will light up, if we make RB5 =0 the LED will turn off.

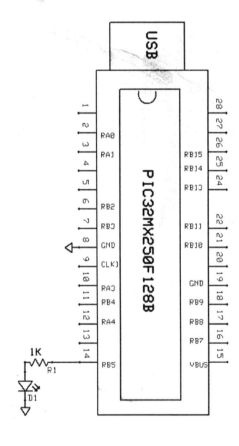

Figure 7-2: Schematic for Timer Experiments

Make sure that the ground potential for your prototype circuit is the same ground reference used by the Microstick II, by connecting the solderless breadboard ground to pin 8 of the Microstick II.

An assembled circuit is shown in Figure 7-3. Notice that ground is picked up from pin 8 and connected to a common bus for ground on the prototype board. This same ground is used for the LED cathode. The cathode must be grounded for the LED to work. You need to be able to discern anode from cathode on the LED to be able to hook it up correctly. The cathode is typically the shorter lead on the device and is also marked by a flat edge on the part.

Figure 7-3: Assembled Circuits for Timer Experiments

Timer 1 Experiment –Non-Interrupt

There is a code snippet from main.c of Timer exercise which shows how Timer 2 is initialized. We let the compiler figure out the PR2 for Timer 2 as follows:

```
#include "system.h"
#define SYS_FREQ                  (8000000L) //8 Mhz internal fast oscillator
// Let compile time pre-processor calculate the PR1 (period)
#define PB_DIV                    1
#define PRESCALE                  256
#define TOGGLES_PER_SEC           2
#define T2_TICK                   (SYS_FREQ/PB_DIV/PRESCALE/TOGGLES_PER_SEC)
```

Figure 7-4: Time Delay Calculation

System.h was discussed in earlier chapters. It contains the configuration pragma to select 8 MHZ FRC internal oscillator as System Clock and set the Peripheral clock to the same value using divide by 1 on System Clock. These facts are copied here to help in the PR2 calculation. T2_TICK is used as PR2 and calculates the necessary PR2 value to ensure that the TOGGLES_PER_SEC value is 2 per second or 500 msec.

This time delay is sufficient for a number of application uses. With this code the time delay is easily set by just changing TOGGLES_PER_SEC value.

```
mPORTBClearBits(BIT_5);                  //Clear bits to ensure light is off.
mPORTBSetPinsDigitalOut(BIT_5);          //Set port as output
  OpenTimer2(T2_ON | T2_SOURCE_INT |T2_PS_1_256, T2_TICK);
  mT2ClearIntFlag();
```

Figure 7-5: Timer and LED port Initialization

Timer initialization is straightforward using peripheral library function

OpenTimer2 (T2_ON I T2_SOURCE_INT IT2_PS_1_256, T2_TICK);

In this function we set Timer 2 SFR setting: PR2 to T2_TICK, set the prescaler TCPS to 256, set TCS source to internal, and turn on Timer 2 T1CON = ON. To ensure T2 flag is cleared we use the library function

mT2ClearIntFlag ();

This approach will work for all the PIC32MX timers. Here a snapshot of the main code. It is broken down into four discrete sections Initialize I/O

- Calculate T2_TICK for delay

- Initialize I/O and Timer 2 using library functions

- o Initializes analog/digital compliment for pin settings
- o Sets SFR T2CON TCPS, TCS, and ON. Sets SFR PR2 for T2_TICK
- o Clear SFR IFS0 T2IF
- Main Loop
  - o Wait on time out (timer flag set) Test IFS0 T2IF
  - o If set Toggle LED and clear flag. Note that is flag is not cleared Led toggling on timeout will not occur.

The important point here is that the timer as used replaces the original delay software functions of earlier chapters. However, in this present design we are still waiting on a timing condition (timer comparison math and T2IF =1) before continuing. A more efficient way that allows for better use of processing resources is required–we will see this using interrupts.

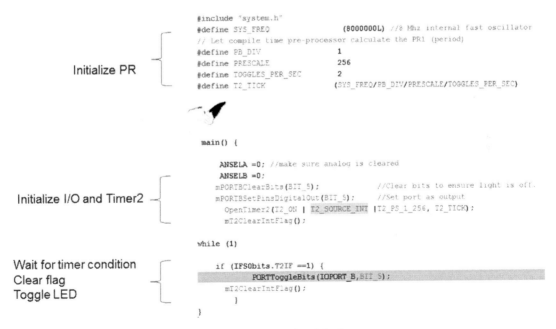

Figure 7-6: Timer Operational Code

Timer 2 - 16 bit Execution

Ok, the hardware is in place, lets discuss the software. Navigate to the MPLAB X project icon Timer1 in Chapter 7 folder first experiment. Let's go ahead and begin debug. Connect the Microstick II. Double click the project icon to bring up the total workspace. At this point we have a fairly good idea of what will happen during execution given that the prototype hardware is configured correctly, and the source code and processor configuration are in good shape.

The PIC32MX will blink the LED connected to pin 14 RB5 of the Microstick II. Let's verify the delay time and ensure that it is sufficient to yield a good blink rate. One quick way is to use the simulation tool "stopwatch" to measure this delay time. Under simulation there are some tools to perform detailed analysis; the "Stop Watch" is an example of this. Under "Stop Watch" we can make actual time measurement. Let's start with setting up simulation and then use stopwatch to verify delay time before working with real time debugging and the Microstick II. The whole procedure should be fairly straightforward given the ground we covered in earlier chapters (see Chapter 2).

Set the breakpoint at the LED toggle instruction in main code. Reset the processor and run. The first time through the stopwatch will capture erroneous time –in that it will capture both the initialization time as well as first time through the loop. Let's zero the stopwatch and then hit run. Now the time should be exact and equal to 500 msec. Compare this time to what was captured earlier –they both should be in agreement. Finally change the debugger from simulator to Starter kit. Rebuild the code for this new debugger selection and run without breakpoint the LED should now be blinking.

Timer 2 - Experiment using Interrupt

Interrupts appear in the PIC32MX C code as unique function declarations as shown in the following template using Timer2 (see Figure 7-7). You will need to follow this template in order for the XC32 'C' compiler to recognize the function code as a specific interrupt vector.

```
void __ISR(_TIMER_2_VECTOR, ipl2) Timer2Handler(void)
{
    // clear the interrupt flag
    mT2ClearIntFlag();
    PORTToggleBits(IOPORT_B, BIT_5);

}
```

Figure 7-7: Timer 2 ISR

Interrupt functions uses interrupt vector and parity settings and returns no parameters (note the use of void), and each interrupt source has its own specific function call. The ISR uses pre-defined vector for Timer 2:

   _TIMER_2_VECTOR

It sets priority to 2 using 'ipl2'. The code that executes during an interrupt function is called the "interrupt service routine". General recommendations for writing an interrupt service are; first, to perform the minimal essential processing needed for the service (that is-- get in and out quickly), and, secondly, do not call any other functions during the service. Make sure that the priority set for the peripheral matching the priority used in the ISR declaration, otherwise the interrupt will not work. Finally, an interrupt service must reset the original interrupt flag that initiated the interrupt before exiting.

Let now view the entire main code. Here's a snapshot. It is broken down into five discrete sections for our present discussions. These sections should be compared against the earlier non-interrupts section.

- Calculates T2_TICK for delay
- Initialize I/O and Timer2 (using interrupt) with library functions
    - Initializes analog/digital compliment for pin settings
    - Sets SFR T2CON TCPS, TCS, and ON. Sets SFR PR2 for T2_TICK
    - Clear SFR IFS0 T2IF
    - Enables interrupt by Setting SFR IEC0 T2IE bit high (turn on interrupt), and setting Interrupt priority to 2 by setting SFR IPC2 T2IP =2
- Interrupt Service Routine-done with library functions
    - Toggle LED
    - Clears flag (T2IF)
- Main Loop
    - Do –nothing main loop –uses Nop () function

Navigate to the MPLAB X project icon Timer1 Interrupt in Chapter 6 experiments folder. If other projects are open in MPLAB X then go ahead and make this project the main project (right click on project name in the project pan and then select pull down option to make it the main project). Let's go ahead and begin debug. Connect the Microstick II. Double click the project icon to bring up the total workspace. At this point we have a fairly good idea of what will happen during execution given that the prototype hardware is configured correctly.

The important point here is that the timer, as used, is again replacing the original delay software functions, however, in this present design we are using timer2 under interrupts. When Timer2 rolls over from PR2 count to zero the condition (T2IF =1) occurs and causes an immediate interrupt in processing, forcing the CPU to respond. The main processing loop actually does nothing in this application, as the interrupt itself is causing the LED to blink.

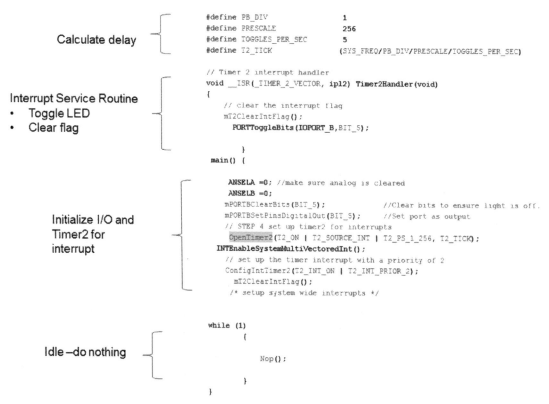

Figure 7-8: Timer 2 Interrupt code examples

The 32 bit Timer Using Timer Peripheral Pairs

A timer high level block diagram of the 32 bit timer is shown in Figure 7-9. This block diagram uses a fixed pair of type B 16 bit timers: Timer 2 and Timer 3 pair and/or Timer 4 and Timer 5 pair. The block simplifies the overall peripheral representation 2 and 4 are designated X, while 3 and 5 are designated Y. The X represents the Least Significant Word (LSW) and Y represents the Most Significant Word (MSW) for the 32 bit operation.

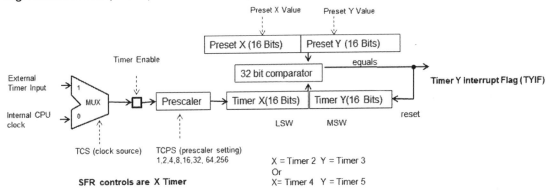

Figure 7-9: 32 bit Timer Block Diagram

Remember there are a total of four type B peripherals in the PIC32MX so for 32 bit operations we can only have two 32 bit timers total. Operations for all 32 timers are similar. You control the setting through Timer X SFR, initialize both PRX for LSW and PRY for MSW, and respond to TYIF for timing condition.

T2_TICK is used as PR2/PR3 (32 bits) and calculates the necessary PR2/PR3 value to ensure that the TOGGLES_PER_SEC value is 1 per second. The T2_TICK is essentially a 32 bit value for a much longer delay then with a single timer. To keep things simple will we keep the delay to a second.

The Timer23 setup is shown (Figure 6-10). Note that only the timer 2 in the pair needs to be set, but both the timer 2 and 3 presets are initialized with the library function as a 32 bit word.

OpenTimer23 (T2_ON I T2_SOURCE_INT IT2_PS_1_256, T2_TICK);

Timer2/3 32 Bit Operation –Non Interrupt

This is a similar exercise to the individual 16 bit non interrupt operation we saw earlier just using Timer 1. Here however we are using Timer 2 and Timer 3 pairs for 32 bit operation. The end objective as earlier stated is to blink a LED. In fact the same hardware configuration is used as per the 16 bit operation. The main source code is shown.

Calculate T2_TICK
```
#include "system.h"
#define SYS_FREQ              (8000000L)  //8 Mhz internal fast oscillator
// Let compile time pre-processor calculate the PR1 (period)
#define PB_DIV       1
#define PRESCALE     256
#define TOGGLES_PER_SEC   1
#define T2_TICK      (SYS_FREQ/PB_DIV/PRESCALE/TOGGLES_PER_SEC)
```

```
main() {
```

Initialize I/O and Timer2/3
```
    ANSELA =0; //make sure analog is cleared
    ANSELB =0;
    mPORTBClearBits(BIT_5);            //Clear bits to ensure light is off.
    mPORTBSetPinsDigitalOut(BIT_5);    //Set port as output

    OpenTimer23(T2_ON | T2_SOURCE_INT |T2_PS_1_256, T2_TICK);
```

Wait for timer condition
Clear flag
Toggle LED
```
    while (1)
    {
      while( !IFS0bits.T3IF );
         PORTToggleBits(IOPORT_B,BIT_5);
         IFS0bits.T3IF =0;
    }
}
```

Figure 7-10: 32 bit Non-Interrupt Exercise

Ok, the hardware is in place, lets discuss the software. Navigate to the MPLAB X project icon Timer23 in Chapter 7 folder first experiment. Let's go ahead and begin debug. Connect the Microstick II. Double click the project icon to bring up the total workspace. At this point we have a fairly good idea of what will happen during execution the LED should blink at a 1 second rate.

Timer2/3 32 Bit Operation Interrupt

This is a similar exercise to the individual 32 bit non interrupt operation we saw earlier but in this case we will be using interrupts. The end objective as earlier is to blink a LED. In fact the same hardware configuration is used as per all earlier exercises. The main source code is shown with a critical section highlighted. Interrupt functions uses interrupt vector and parity settings and returns no parameters (note the use of void), and each interrupt source has its own specific function call. The ISR uses pre-defined vector for Timer 23, _TIMER_23_VECTOR and it sets priority to 2 using 'ipl2'.

Navigate to the timer23 with interrupt MPLAB X project file in Chapter 7 folder. Let's go ahead and begin our debug. Connect the Microstick II. Double click the project icon to bring up the total workspace. At this point we have a fairly good idea of what will happen during execution. The PIC32MX will blink the LED connected to pin 14 RB5 of the Microstick II.

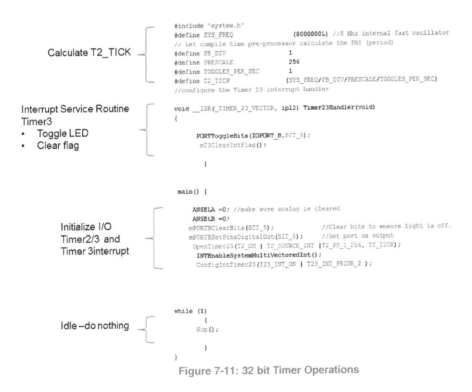

Figure 7-11: 32 bit Timer Operations

Review of PIC32MX Timers capability:
- Five 16 bit general purpose timers/counters
  - Similar functionality between all 5 timers
  - Period registers for each
  - Interrupt generation on match with period
  - Reset on match
- The timers are broadly classified into two types, namely:

    Type A: (Timer 1)
    - Asynchronous timer/counter with a built-in oscillator
    - Operational during CPU Sleep mode
    - Software selectable prescalers 1:1, 1:8, 1:64 and 1:256

    Type B: (Timers 2, 3, 4, 5)
    - Ability to form a 32-bit timer/counter
    - Software prescalers 1:1, 1:2, 1:4, 1:8, 1:16, 1:32, 1:64 and 1:256
    - Event trigger capability (Timer 3 only ADC event trigger)

- Four of these timers B type (Timer 2+3 and 4+5) can make two 32 bit timers/counters

- In 32 bit operations :Timer 2+3 or Timer 4+5= Least Significant Word and Most Significant Word

- In 32 bit operations: Use SFR of first timer for control, initialize both Presets, and then use second timer for interrupt operations

Review of Source Code Key Features:
- Two new ISR vectors were used for Timer 2 and Timrer23
- Similar library functions were used to set up the timers.

    OpenTimer2 (T2_ON | T2_SOURCE_INT |T2_PS_1_256, T2_TICK);
    OpenTimer23 (T2_ON | T2_SOURCE_INT |T2_PS_1_256, T2_TICK);

- The only different is that the PR size has increased from 16 bits to 32 bits
- For 32 bit operation you set up using Timer 2 SFR and clear Timer 3 Interrupt flag.

Exercise:
1. Verify time delays using simulation, breakpoint, and stopwatch tools for Timer1 interrupt and Timer2/3 interrupt

2. Try different values of preset in experiment. Perform the expected delay calculation and then verify.

3. Changing the first experiment using another timer –say timer 2.

4. Change the third experiment by using timers 4/5—use different presets, calculate delay and verify by simulation.

5. Change the fourth experiment by using timers 4/5.

Chapter 8 - Optimizing PIC32 Performance

In this chapter we will explore techniques to enhance PIC32 performance. We will work through a series of exercises, performing benchmarks and examining the impact of three C's techniques to make an application run faster.

1. Cache and Wait State **C**onfiguration
2. System **C**lock
3. 'C' **C**ompiler Code Optimization

Let's start with configuration. We will first detail the MIPS 32 4K Core operations and how it functions in the system and the role of pre-fetch cache in FLASH Memory retrieval and RAM.

MIPS 32 4K Core Instruction and Data Operations

The PIC32 is based upon the MIPS 32 4K core. The MIPS is a leading technology for high performance cost sensitive high volume microcontrollers. The MIPS 32 4K performance has been benchmarked at an impressive 1.56 DMIPS (Dhrystone Millions of Instruction per Second) per MHz. The Dhrystone represents a standard computing benchmark based upon IBM 11780 computers. PIC32 is one the highest performance 32 bit devices in its class.

The PIC32 block diagram (see Figure 8-1) highlights the MIPS 32 4K core and the internals of the chip organized around two internal buses. The top bus connects the Bus Masters: The MIPS 4K core, 4 channels DMA (Direct Memory Access Controller), USB Controller, and other devices. The other devices are pre-fetch cache, ram, interrupt controller and all the digital ports and a peripheral bridge to all the on–chip peripherals. The top bus is the fastest and runs at the system CPU clock rate. It is not so much a bus but in reality a high speed switch matrix. As a switch matrix it allows simultaneous communications for bus masters on the bus and other devices hooked to the bus without any contention. The other bus, the Peripheral Bus, is slower, and can be programmed to run at a different clock rate than the CPU. The exact Bus clock is determined by the Peripheral Bridge setting.

The MIPS core is based upon a Harvard Architecture using separate data and instructions paths (see the two distinct paths into the core from the bus matrix). Flash memory in this case can't run as fast as the CPU. To further enhance the performance; the PIC32 employs a 128-bit Pre-fetch Cache module and PIC32 a 128-bit wide Flash memory. Such a wide memory path is specifically designed to increase the instruction throughput and improve overall CPU performance. This module can be programmed to look ahead and pre-fetch the next 128-bits of instructions and store them in an on-chip cache memory. This allows programmed flash to keep up the high speed MIPS 4K core achieving an instruction execution per clock pulse once the internal pipeline has filled.

Another feature to help with performance is that Core execution can occur from either high speed RAM or Flash. This is because the PIC32 uses the unified memory map – meaning that both Instruction and Data space reside in one linear address space, each occupying a unique range of addresses. RAM execution is done using a compiler directive during build. However, RAM implementation in a microcontroller requires larger chip real estate so RAM is at a premium where Flash sizes are typically larger.

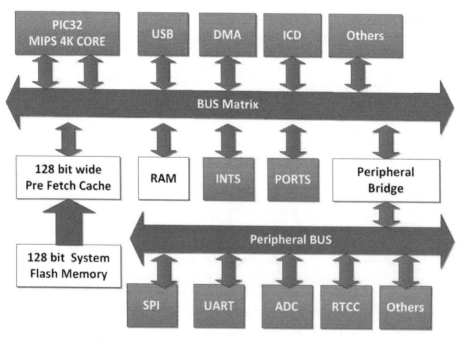

Figure 8-1: PIC32 Block Diagram showing MIPS Core

MIPS 32 4K Internal Operations

The MIPS 32 4K core uses a 5-stage execution pipeline. This means that each instruction is executed in 5 different stages. Once the pipeline is full, the M4K core executes one instruction per CPU clock. The PIC32 uses the high-performance version of the Multiply and Divide hardware module. A very powerful feature of this module is that it contains its own autonomous pipeline. It takes 1 cycle to perform 16x16 or 32x16 multiply operations, and 2 cycles for other sizes. The divide operation takes from 11 to 32 cycles. Exact cycle count depends on the dividend operand size. The PIC32 executes 32-bit instructions. The 32-bit instructions are designed to provide higher performance. If the application is code size sensitive, it may use MIPS16e™ instructions. The MIPS16e instructions are 16-bit wide. With the use of MIPS16e instructions, applications can save up to 40% of code size compared to the 32-bit instructions (see Figure 8-2).

The PIC32MX core execution unit implements a load/ store architecture with single-cycle ALU operations (logical, shift, add, subtract) and an autonomous multiply/divide unit. The PIC32MX core contains thirty-two 32-bit general-purpose registers used for integer operations and address calculation. One additional register file shadow set (containing thirty-two registers) is added to minimize context switching overhead during interrupt/exception processing. The register file consists of two read ports and one write port and is fully bypassed to minimize operation latency in the pipeline.

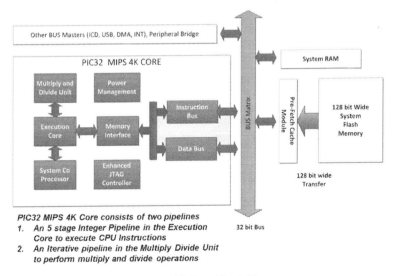

Figure 8-2: MIPS Core Block Diagram

There are five stages associated with Execution Core pipeline (reference Figure 8-3).

1) Instruction Fetch
2) Execution
3) Memory Fetch
4) Memory Align
5) Memory Write back

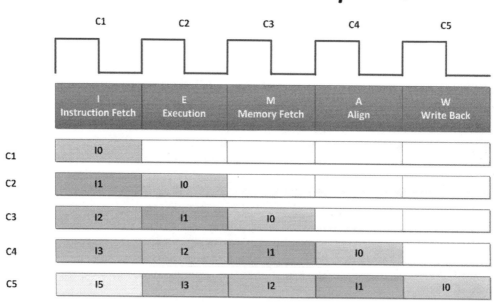

Figure 8-3: MIPS Core Execution Pipeline

As the name suggests, in the Instruction fetch, an instruction is fetched. In stage E, the instruction is decoded and executed. In the M stage, memory operands are fetched from the on-chip memory – be it SRAM or Flash. In the A stage, the memory data is word aligned and in the W stage the result is written to the destination.

First Optimization- Cache and Wait State Configuration

The cache is configurable and can be bypassed at lower CPU Clock rates. Flash Memory wait states are also configurable based upon size of the Flash and CPU clock rate. RAM can work at near full rate with MIPS core but when CPU clock exceeds 40MHZ a single wait state is required.

Core to memory optimization is achieved with specific library calls that balance Clock, Waits states and Pre-Fetch cache configuration.

One such library call is:

SYSTEMConfig (SYSTEM_CLOCK, SYS_CFG_WAIT_STATES | SYS_CFG_PCACHE);

We will examine its use in the exercises to follow.

Second Optimization – System Clock

Clock rates and sources are completely programmable with the PIC32MX. A simplified block diagram illustrating clock sources for System Clock (Bus Matrix and CPU clock) and Peripheral Clock (Peripheral Bus and all chip peripherals) is shown (Figure 8-4). The Microstick II has no external clock crystal, so only the internal Fast RC Oscillator (FRC) will be used as the source. All the blocks and sources are configuration bits defined in Figure 5. The combinations are a little daunting.

There is a tool within MPLAB X that allows us to view and select configuration (see Figure 8-6 steps #1, #2, and #3). As a fourth step once we selected the configuration we can then have MPLAB X automatically generate the code. This code is then cut and pasted in our code and appears as pragma settings. In our case the configuration code resides in project file System.h. For all cases we need to insure that clock configuration does not exceed the 40 MHZ max clock rating for the PICMX250F128B part. In our exercises we will cover two system clock cases once at 4MHZ and the other at 36MHZ. Both are derived from the Internal FRC of 8 MHZ. We will ignore peripheral clock since it will not be used as part of the exercises.

Exercise 1:
System.h snippet looks like the following. We are configuring system clock to be FRC divided. The division will occur in the main code and will be by 2 for 4MHz.

// DEVCFG1
#pragma config FNOSC = FRCDIV   // Oscillator Selection Bits
                                // (Fast RC Osc w/Div-by-N (FRCDIV))

Exercise 2:
System.h snippet looks like the following. We are configuring system clock, using FRC through PLL (Phase Lock Loop) to be initially input divided by 4, then multiplied by 18, and then output divided by 1. The end result is 36MHZ or 8MHZ/4*16*1.

#pragma config FPLLIDIV = DIV_4   // PLL Input Divider (4x Divider)
#pragma config FPLLMUL = MUL_18   // PLL Multiplier (20x Multiplier)
#pragma config FPLLODIV = DIV_1   // System PLL Output Clock Divider
                                      // (PLL Divide by 1)

// DEVCFG1
#pragma config FNOSC = FRCPLL   // Oscillator Selection Bits (Fast RC
                                    // Osc with PLL)

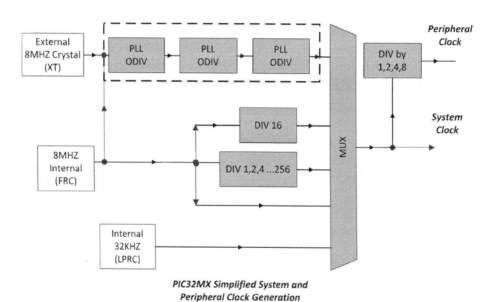

PIC32MX Simplified System and Peripheral Clock Generation

Figure 8-4: Simple System and Peripheral Clock Sources Block Diagram

Figure 8-5: Clock Configuration bits

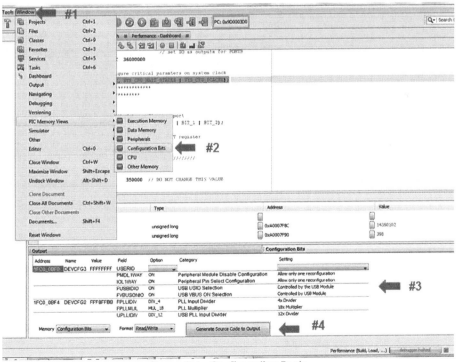

Figure 8-6: Configuration Tools

Beginner's Guide to Programming the PIC32 　　　Page 155

Third Optimization – Compiler Settings

The X32 C Compiler has optimization options that can be set during compile time to enhance code execution. You access this tool through the dashboard using the icon below.

You select the complier XC32-GCC and select optimization category. The different levels can be set. Unfortunately for most, the complier is a free evaluation type and the optimization level is fixed. So the impact of this will not be covered in the benchmarks.

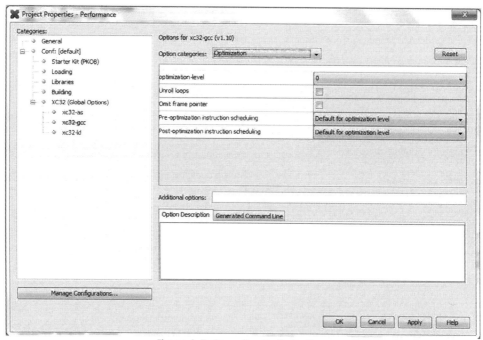

Figure 8-7: Compiler Optimizations

Exercises

Build a single prototype as per Figure 8-8 schematic. The exercises are very similar in nature with the only real different is clock setting (as described earlier). The same prototype will be used in both exercises. Three LEDs are used all have a series 1K resistor to ground. LED1 is connected to RA0 pin 2, LED 2 is connected to RA1 pin 3 and LED 3 is connected to RA3 pin 9.

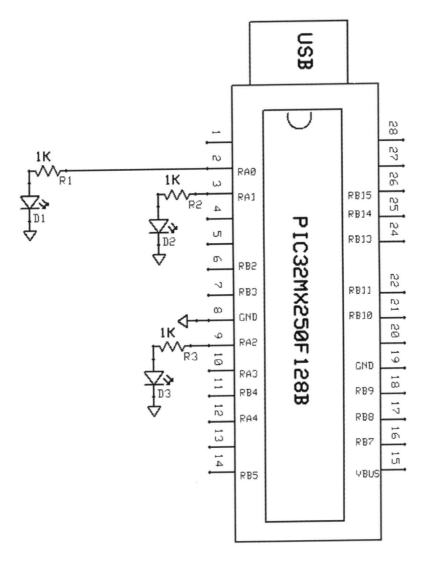

Figure 8-8: Schematic

The LEDs will serve as a display to show some activity during long loop times involved in the benchmarks.
The long delay is given by the follow snippet.

```
// create a delay and display on LEDS
    #define DELAY    350000  // DO NOT CHANGE THIS VALUE
    for(temp=0; temp < DELAY; temp++);
```

```
        for(temp=0; temp < DELAY; temp++)
              PORTAINV=7;
        for(i=0; i < DELAY; i++)
        {
              if (((i>>17)&7) != previous)
              {
                    led_disp(i);
                    previous = (i>>17)&7;
              }
        }

void led_disp(int val)
{
        mPORTAWrite( val>>17);
}
```

Prior to the delay we use core timer library function to initialize the benchmark clock

```
// C32 macro to clear Core Timer COUNT register
    _CP0_SET_COUNT(0);
```

At the end of delay prior to a final breakpoint we will tally up the time in milliseconds and CPU required for executing the delay shown above. This delay will be conducted for different clock settings and will have optimization turned on and off and comparison made.

```
// C32 macro to read Core Timer COUNT register
    ticks = _CP0_GET_COUNT();

  cpu_ticks = ticks * 2;  // cpu_ticks is a long integer
  time = cpu_ticks/(SYSTEM_CLOCK/1000); //time is captured in
                                        // milliseconds

    While (1); // this is where the breakpoint is set.
```

Both time and cpu_ticks are set as watch variables and display in decimal. Open MPLABX and navigate to folder excercise1 performance for 4MHz clock operation. Open project exercise 1. It is a project with two files System.h and main.c configured for 4 Mhz operation.

A listing main.c is shown in Figure 8-9 illustrating benchmark application.

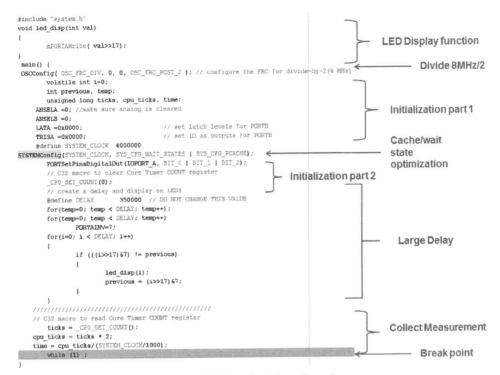

Figure 8-9: Exercise 1 benchmark

Connect Microstick II, and then build debug project. Make sure break point is set at the final while statement in main.c that and watch variables time and cpu_ticks are visible. Run with optimization statement comment out and uncomment. These should result in two sets of readings. The LEDs should flicker during delay. Any issues should be the result of wiring or Microchip II placement on prototype board. Check and redo. Record readings and then close project.

Open MPLABX and navigate to folder exercise 2 for 36 MHz clock performance. Open project exercise 2. Build debug project. Again run two times with optimization statement commented out and then uncommented. Record reading and then close project. We should now have 2 sets of reading one for 4MHZ clock and the other for 36 MHz clock as follows:

- At 4MHZ (no optimization) time = 4550 msec cpu_ticks = 18,200,130
- At 4MHZ (optimization)   time = 3587 msec cpu_ticks = 14,350,102
- At 36 MHZ (no optimization) time = 505 msec cpu_ticks = 18,200,130
- At 36 MHZ (optimization)   time = 398 msec cpu_ticks = 14,350,102

It should be no surprise that performance measured in elapsed milliseconds improved with increasing clock rate, optimized or not, the Clock ratio of 36 to 4 results in a 9X improvement in performance.
In applications where clock cannot be increased optimization alone (allowing the system to adjust cache and waits states for given clock), yields a reduction total 3,850,028 cpu_ticks required and achieves 20% improvement in performance.

Review of Enhancement Techniques:

- There are three general techniques for improvement PIC32 application performance. They are associated with

    - Configuration
    - Clock
    - Compiler

- Configuration is associated with optimizing the MIP 32 4K core interaction with program flash and RAM for a given system clock. In our examples this yielded a 20% increase in performance.

- Clock rate increase is an expedient way to enhance performance. The PIC32 has numerous options to select and set system clock through configuration setting. These setting appear as pragma directives in your source code. MPLAB X has utility features to help.

- Make sure the system clock, as configured, does not exceed the maximum clock rate for your part. In the case of the PIC32MX250F128B that rate is 40MHZ.

- The fully licensed XC32 C Complier has optimization levels that can be set for compile. When used, this should also yield performance increases.

Review of Source Code Key Features:

- Core to memory optimization is achieved with specific library calls that balance Clock, Waits states and Pre-Fetch cache configurations. Once such library call is

SYSTEMConfig (SYSTEM_CLOCK, SYS_CFG_WAIT_STATES | SYS_CFG_PCACHE);

- Use #pragma config FNOSC = FRCDIV to divide down 8MHZ FRC. Perform final division as part of main code
- Use #pragma config FNOSC = FRCPLL to multiply up 8MHZ FRC

Exercise :

4. Try redoing exercises with a different clock rate.
5. Test benchmark with longer delays.
6. What are the 3 C's?
7. Explain basic core 5 step pipeline
8. How does the MIPS 32 4K core interface to Flash Memory in the PIC32
9. At what rate does the Bus matrix run?

Chapter 9 – Serial RS232 Communications

In this chapter we will introduce PIC32MX serial communications using a UART (Universal Asynchronous Receiver Transmitter) and then add this capability to our applications library. We will also introduce the Microchip configurable peripheral pin assignment system designated Peripheral Pin Select (PPS) to configure the PIC32MX output pins for the UART2 peripheral. Our examples will cover interrupt and non-interrupt operations for RS-232 and USB interfaces to a PC. Let's begin with a general overview of serial communications.

Serial Communications Overview

In general, serial communications is microcontroller communication that is performed by transmitting and receiving data bytes one bit at a time. With the UART we introduce a type of serial communications that is called "asynchronous", because there will no shared clock is used in the communications. If a shared clock is used then the serial communications is termed "synchronous".

In asynchronous communications the transferred bits are composed of not only data but also framing bits for start, stop and error checking. This communication is one of the oldest legacy types used by microcontrollers and still very much actively applied. The communication can exist between microcontrollers, or a microcontroller to a PC or display, a microcontroller to digital camera, or even a wireless transceiver. Within each communicating device there is a UART or equivalent.

The PIC32MX has two completely independent UARTs; UART1 and UART2. For our experiments we use UART2 to communicate with a PC. The PC already supports serial asynchronous communications with both software applications (TeraTerm and HyperTerminal) and hardware interfaces (USB and RS-232 port). Each PIC32MX UART peripheral has separated transmit and receive hardware.

The UART 2 Transmit register (U2TXREG) contains the data byte to be transferred, and it functions as a PISO (Parallel in Serial Out) digital register. The transmit section takes in a parallel data (byte) from the PIC32MX and transmits it as a serial bit by bit stream at a fixed bit or baud rate. The receive section, captures incoming data from an outside device and functions as a SIPO (Serial in

Parallel Out) digital register. The UART2 Receive register (U2RXREG) takes in serial data and forms a parallel data (byte) for transfer to the PIC32MX CPU.

The complete UART peripheral contains not only these functions but also an internal bit clock generator, framing logic to remove the data byte from the overall data stream, and additional receive data buffers to prevent data overflow. For the PIC32MX UART the baud rate clock is generated by configuring an internal clock source that is derived from the peripheral bus clock.

A basic requirement for asynchronous communications to work is that each (UART) is configured with the identical bit rate clock or baud rate clock, and framing; the start bit, the size of data (7 or 8 bits), the number of stop bits (1 or 1.5) and finally, if there is error correction, the type of parity used (even or odd). Another basic requirement is that the UARTs are "wired" together correctly. What this means is that each UART TX side be connected to the other UART RX side so data can flow between them. These connections can be Wired (direct physical connection) or Wireless if (radios are used).

During communications the UART transmitter side alerts the receiving device UART that a byte is about to be transferred using framing. It does this by lowering the serial data line for a bit time (designated the start bit). The receiver UART uses this start bit to synch up its internal baud rate generator to this start bit to begin sampling of the incoming data. Data is then transferred one bit at a time for the 8 data bits that make up the data byte and then an error correction or parity bit can be sent (this is optional), and finally a stop bit is a fixed high for one or two bits times (stop bit size is optional as well). This leaves the serial data line is a high condition for the start of the next frame. See Figure 9-1 below.

This leads us to more definitions. When only one UART is used to transmit its data traffic at a time the communications is designated "half-duplex". However, when the configuration allows for both UARTs to simultaneously transmit data, the communication is termed "full-duplex".

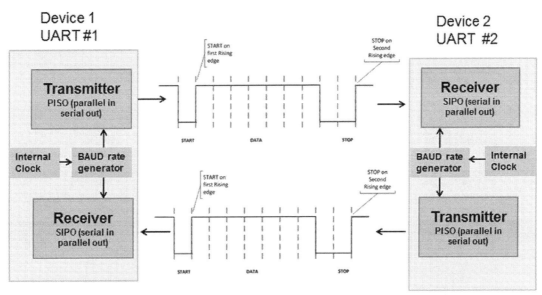

Figure 9-1: Basic Serial Asynchronous Communication

Several practical examples of microcontroller serial communications with different peripheral types are shown. In all these cases both the microcontroller and peripheral reside in the same equipment and serial communication is used as a way of internally connecting the microcontroller and peripheral for command, control, data transfer, and status. The peripheral may be the end point in the communication, like the display and voice recognition modules, or may serve as a bridging device to outside communications with a remote device, like the WIFI and ZigBee wireless modules.

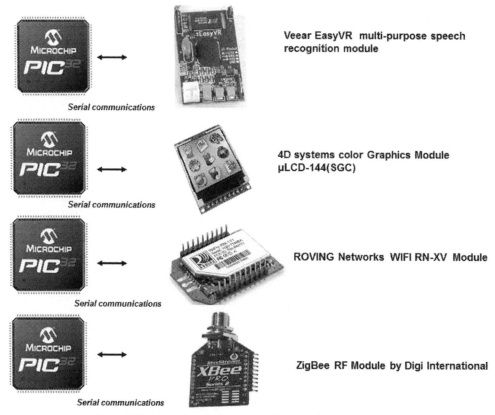

Figure 9-2: Examples of Microcontroller serial communications

In wireless serial communications transmitters typically share the same carrier frequency forcing the communications to be half-duplex (only one transmitter on at a time). This helps insure that the serial communication avoids data collisions (where two sides try communication at the same time, resulting transmission signal interference). Coordination of wireless communication is done typically with use of a "higher layer" software protocol (like 802.15 ZIGBEE or 802.11 WIFI) that resides over this "lower layer" data link to establish who can talk and when.

Industry Standard External Wired Serial Communications

There are several legacy industry standards for externally connecting wired connecting of equipment's that have UARTs. One of these standards for point to point communication is RS-232. The standard dictates the connector types, baud rates, pin outs and voltage levels. RS-232 requires that the voltage level from +3.3V to 0V of the PIC32MX UART output must be modified to -12V to +12V to meet standard. Voltage level changing electronics is needed.

The RS-232 standard also uses a common connector set 9 pin DIN male and female connector pair. For RS-232 compliance one side must be configured as DCE (data communications equipment) and the other DTE (data terminal equipment). This standard partitioning guarantees each side's transmitted and receives pin outs and connector types work to make communications possible. The PC is pre-configured as DTE. This requires the PIC32MX or Microstick II to be DCE. RS-232 also supports hardware controlled data flow with CTS (Clear to send), DSR (Data Set ready), and RTS (Ready to Send) signals between end devices to facilitate data transfer. This feature is optional.

A complete end to end hookup diagram for a Microstick II connection to a PC using RS-232 is shown in Figure 9-3. The PC runs a software application program (Tera Term or HyperTerminal) to allow the serial data communications stream to be visible on the PC screen. HyperTerminal is a "for sale" application, and Tera Term is free open source. These apps will be used in our experiments. For the level conversion (+3.3/0 VDC to -12VDC/+12VDC) hardware and a DCE connector for RS-232 for the Microstick II we use an ACRONAME module. It contains both level conversion and interface in a convenient pluggable board. We will then use a standard RS-232 cable DCE to DTE pair to connect the PC to brainstem.

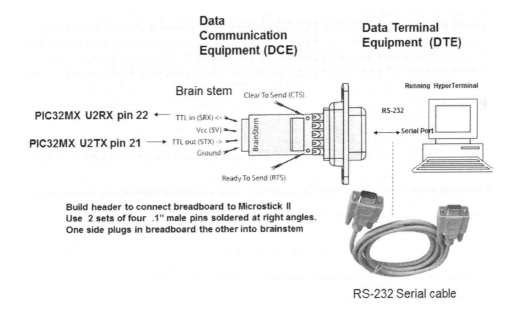

Figure 9-3: Example PC Serial RS-232 Communications Scheme

As a reminder the data rates and serial format for serial communication needs to be fixed on both side. This is necessary for the PIC32MX UART initialization software, and HyperTerminal/Tera Term configuration. The settings are baud rate, number of stop bits, data size, use of parity, and finally any hardware handshaking. For our experiments we will use 2400 baud, 8 bits, no parity, 1 stop bit and no hardware flow control. The Short hand notation for this is 2400N81

Another legacy serial communication industry standard is RS-485. This is used for connecting a full network of equipment serially. It is the predominant in industrial and military equipment monitoring and control. RS-485 is half-duplex and a low level software protocol is used where one equipment is designated as a bus master and resides over the shared communications bus determining when other equipment (slaves) can talk. Each entity on the bus has a unique 9 bit address that is used to facilitate this communication.

The PIC32MX UART has this extended capability. The voltage levels here are differential (this allows for longer wire length (up to 30 meters) then the previous RS232 (10 meter limitation between equipment) and, again, a voltage level change is necessary to accommodate +5 volt and -5 volts differential signals using +3.3 /0 volt from PIC32MX UART.

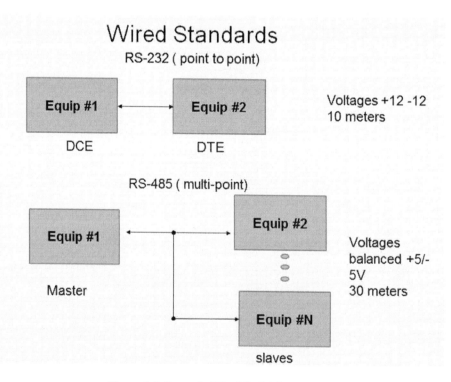

Figure 9-4: Example Wired Serial Standards

USB (Universal serial Bus) to PC Connectivity

USB or Universal Serial Bus was co-developed by a number of large companies, including Microsoft and Intel with the explicit purpose of making it easier to add/remove peripheral devices from a PC. If you remember not too far back in time, the older generation of PC device peripherals had their own special purpose RS-232 serial communication ports, parallel printer ports, and PS/2 for keyboards. Today, all of that is gone with USB. USB 1.1 release in 1998, followed by USB 2.0 release in 2000 has become the defacto standard for PC peripheral device interfaces. To address the particular needs for a large number of device peripheral devices, USB introduced the concept of "Device Classes".

These are:

- Human Interface Device Class (HID) – joystick, mouse, keyboard
- Mass Storage Device Class (MSD) – external hard drives
- Communication Device Class (CDC) – modems, Ethernet adapters

The class of interest to us is CDC. We will use a UART to USB Bridge that allows for a UART interface on microcontroller side and USB CDC compliant device interface on the PC side. ACRONAME provides a pluggable module with their S27. The S27 module provides the bridge chip with the same four pin connector for the micro side and a USB device connector on the other side. You connect to the PC using the familiar USB Host type A to Device Type B connector cable. USB supports "plug and play".

Each device, when it is connected to the USB, is "enumerated" by the PC, which means the device identifies to the PC its class type, operational needs, and driver requirements. The PC checks that it can meet the operational needs and that the identified driver for the device is present in its current software registry. Only when all this is accomplished, will communications be allowed. We can then open Tera Term on the PC and identify the COM port associated with the S27, configure the COM port, and begin communications.

We will use the ACRONAME supplied registry file (.INF) to register the communication port with the Operating System and the ACRONAME supplied S27 driver. When first connecting to the PC, the system will see the S27 as "new hardware" on the USB and will attempt to retrieve driver information. At this point you direct the PC to the ACRONAME driver locations you stored on your PC. This is a one-time configuration requirement which will then afterwards be connected automatically every time you connect again. The USB is +5V bus, however, we only need to power up the S27 with available +3.3V on the micro USB (Universal serial Bus).

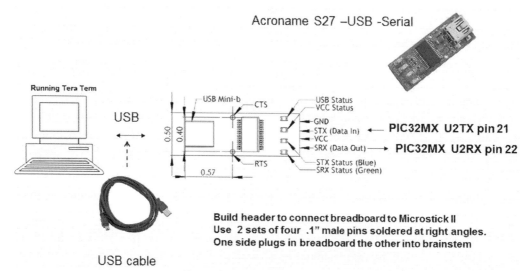

Figure 9-5: USB communications

American Standard Code for Information Interchange (ASCII)

In reality, a user can select any type of digital byte for transfer but HyperTerminal and TeraTerm can only display ASCII (American Standard Code for Information Interchange) formatted data. A table is shown in Figure 9-6. The table is for 128 different characters that includes control, data (upper /lower case A-Z), numerical (0-9), special characters like quotes, exclamation point, question mark, and pound sign. We will be using ASCII data coding for our serial communication experiments. So ASCII data will make up all of our byte data communication content.

b_4	b_3	b_2	b_1	Row	0	1	2	3	4	5	6	7
0	0	0	0	0	NUL	DLE	SP	0	@	P	`	p
0	0	0	1	1	SOH	DC1	!	1	A	Q	a	q
0	0	1	0	2	STX	DC2	"	2	B	R	b	r
0	0	1	1	3	ETX	DC3	#	3	C	S	c	s
0	1	0	0	4	EOT	DC4	$	4	D	T	d	t
0	1	0	1	5	ENQ	NAK	%	5	E	U	e	u
0	1	1	0	6	ACK	SYN	&	6	F	V	f	v
0	1	1	1	7	BEL	ETB	'	7	G	W	g	w
1	0	0	0	8	BS	CAN	(	8	H	X	h	x
1	0	0	1	9	HT	EM	)	9	I	Y	i	y
1	0	1	0	10	LF	SUB	*	:	J	Z	j	z
1	0	1	1	11	VT	ESC	+	;	K	[	k	{
1	1	0	0	12	FF	FC	,	<	L	\	l		
1	1	0	1	13	CR	GS	-	=	M	]	m	}
1	1	1	0	14	SO	RS	.	>	N	^	n	~
1	1	1	1	15	SI	US	/	?	O	_	o	DEL

Figure 9-6: ASCII table

Beginner's Guide to Programming the PIC32

Serial Communications Review

Some important reminders and rules of thumb with serial communications:

- Message is equal in size to a Start bit, # of stop bits, # of data bits , parity if present (10 bits)
- Experiments will use 1 start 1 stop, 8 data bits, and no parity
- Baud rate
    - Bit transfer rate during message up to 115K
    - Experiments will be using 2400
- Data bits
    - Number of data bits transferred in a message
    - In our experiments 8 bits is used
- Parity
    - Optional Parity error checking bit present in message
    - Parity can be set as None, Even or Odd.
    - In our experiments the parity is none.
- Overall UART Communication
    - 2400 baud ,8bit, no parity , one stop bit or short hand 24008N1
- Hardware Flow Control
    - RS-232 supports hardware flow control (Clear to Send CTS), (Ready to Send RTS) and (Data Set ready (DSR) between receiver and sender to prevent communications overflow between receiver and transmitter. This is in addition to the basic RX, TX and GND message communication lines. Given the rates we are transmitting at and our complete control of the receiver and transmitter, we will not use CTS/DSR handshake features if needed. However the PIC32MX does support this hardware handshake.

- RS-485 uses a special 9th bit for addressing between master and slave –slave can only talk after being addressed

- Voltage level changing electronics is necessary for interoperability with the microcontroller UART and any external serial communication standard.

- For internal control between microcontroller and an associated intelligent peripheral serial communication is often used. Typically in this case voltage leveling electronics is not required.

- USB (Universal Serial Bus) has emerged as the de facto standard for PC communications. There are UART to USB chip sets that bridge the gap for Microcontrollers.

In summary the benefits of serial communications with microcontrollers are listed here.

1. It is a widely used Universal Communication scheme
2. Simple to use
3. Some implementation require only one wire (were microcontroller is talking to a display)
4. Easy to troubleshoot
5. Straightforward to use for connecting a microcontroller to a PC supplied with HyperTerminal Software or Tera Term (more on this later)
6. No special licensing needed.
7. Expandable with USB-to-Serial adapters (again more on this later)

The PIC32MX Peripheral Programming System (PPS)

Before we can use the PIC32MX UART we must understand the PIC32MX PPS system. This system allows users to configure which of a select number of digital peripherals appear on (RP) pins of the PIC32MX device. Our direct interest is to assign UART to PIC32MX pins 22, 21. Gone are the days when pin outputs for a chip are hard-wired. With the new PIC32MX and its smaller packages (like the 28 pin skinny DIP) we can essentially configure our own.

Remember the PIC32MX has a large digital peripheral set within the chip. The PIC32MX250F128B pin layout is shown along with each individual pin assignment is shown in Figure 9-7. If all the internal peripherals are used then a total required pin count of approximately 25 pins are needed as derived from the following list.

- Five 16-bit timers (5 input pins)
- Five input capture –automatic pulse capture and measurement (5 input pins)
- Five Output Compare / PWM modules –automatic pulse generation (5 output pins)
- Two UART (Universal Asynchronous Receiver Transmitter) with address capability (4 pins (2 in and 2 out))
- Two SPI (Synchronous Peripheral Interface) (6 pins (2 data out, 2 data in, 2 clock (in/out))

Obviously the total peripheral pin needs are large, in fact, and with these 28 pin PIC32MX devices, there are more internal peripherals then there are available pins. As a compromise solution the PIC32MX250F128B 28 pin device supports PPS with 14 remappable (RP) pins. These remappable pins are configurable as input or output to any of the above peripheral set. These pins are designated with "RP" for reprogrammable pin. This is the closest thing to layout of your own device.

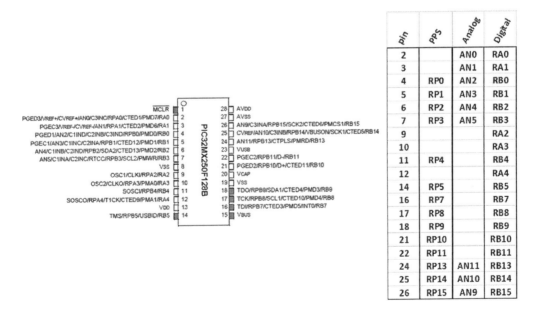

Figure 9-7: PIC32MX250F128B PPS pin set

Initializing PPS

Peripheral Pin Select or PPS has to be initialized, and it can set for only a single configuration to eliminate any changes accidentally during operation. To achieve this Microchip has instituted a specific sequence of steps that need to occur for PPS configuration and a device fuse configuration bit.

Here's the basic idea; at the start of the main program, you can unlock, and configure the PPS for designated peripherals, configuring first the outputs, and then the inputs. Once this is done the PIC32MX is locked and remains with this pin state configuration. You can select reconfiguration options to be one time or many depending using DEVCFG3 setting. What is shown is enabling only a one time setting as currently in the System.h file.

```
// DEVCFG3
#pragma config IOL1WAY = ON      // Peripheral Pin Select Configuration
                                 //(Allow only one reconfiguration)
```

By setting this once, the PPS cannot be undone unless reset occurs or an unlock sequence. The PPS lock and unlock controls are really implemented as special code sequences implemented in the Microchip library as PSS Input () and PSS Output (). This insures that random code reads or writes cannot access and contaminate the fixed PPS settings. Let's review library function help and where you can locate it. The help is located in the XC32 complier folder you installed in Chapter 1. Navigate to it as follows local Disk (C) -> Program Files (x86) -> Microchip -> xc32 -> v.1.10 -> docs. Double click the following icon to take you to Peripheral library help.

Microchip-PIC32MX-Peripheral-Library

You should see the following. A tree structure with all perpherials on left and expanations on the right

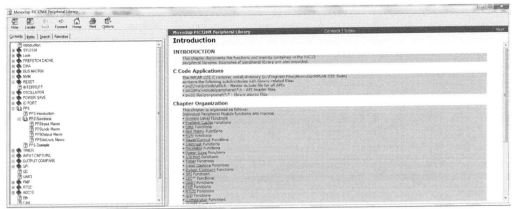

Figure 9-8: Microchip Peripheral Library Help

Navigate to PPS input function and you should see the following, a complete description of the function and its arguments as well as an example. This will serve as our template. You can also similar information on PPS output function.

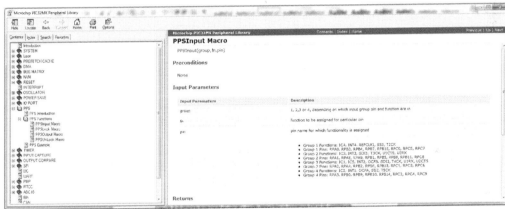

Figure 9-9: PPS Library Function help

By using the library function the required SFR setting and execution of the specific PPS lock and unlocks code sequences happens automatically. Lock and unlock sequences ensure that critical SFR are not inadvertently tampered with during code execution

Figure 9-10 shows an example of library code required to initialize PPS to configure RPB11 for UART2 INPUT receive and then RPB12 for UART2 Output transmit.

```
/*********************************
         main entry point
*********************************/
int main ( void )
{       /*      Initialize PPS */

PPSInput(2, U2RX, RPB11);        //Assign U2RX to pin RPB11
PPSOutput(4,RPB10, U2TX);        //Assign U2TX to pin RPB10

ANSELA =0; //make sure analog is cleared
ANSELB =0;

initU2();
```

Figure 9-10: PPS initialization for UART2

INT3	INT3R	INT3R<3:0>	0000 = RPA1
			0001 = RPB5
T3CK	T3CKR	T3CKR<3:0>	0010 = RPB1
			0011 = RPB11
IC3	IC3R	IC3R<3:0>	0100 = RPB8
			0101 = RPA8
			0110 = RPC8
U1CTS	U1CTSR	U1CTSR<3:0>	0111 = RPA9
			1000 = Reserved
U2RX	U2RXR	U2RXR<3:0>	.
			.
SDI1	SDI1R	SDI1R<3:0>	1111 = Reserved

Figure 9-11: Group 2 Input PPS pin assignments

In this initialization UART2 RX is assigned to RPB11. In general for PPS input assignments you need to verify in the PIC32MX data sheet table (shown partially above) that a PPS connection is possible. The table can be found in the PIC32MX250F128B data set under PPS input assignment. For the library function PSS Input () arguments for this table is defined as group 2, and we are verifying that a connection of RPB11 (or RB11) to U2RX is possible. The data sheet verifies this, so that a functional argument for PSS Input (2, UR2X, RB11) is used. Another way this can be visualized is with the following picture where an input multiplexer for pins to U2RX is controlled. The function sets an SFR that contains the U2RXR bits with the value 0011, enabling the specific pin RB11 to U2RX peripheral connection.

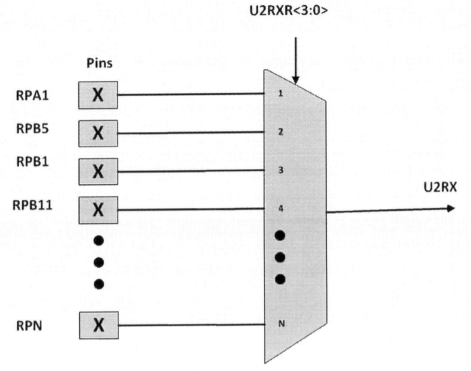

Figure 9-12: Remappable Inputs for UART2 RX

For PPS output the initialization scheme is different.

RPA3	RPA3R	RPA3R<3:0>	0000 = No Connect
RPB14	RPB14R	RPB14R<3:0>	0001 = $\overline{U1RTS}$
			0010 = U2TX
RPB0	RPB0R	RPB0R<3:0>	0011 = Reserved
			0100 = $\overline{SS2}$
RPB10	RPB10R	RPB10R<3:0>	0101 = OC3
RPB9	RPB9R	RPB9R<3:0>	0110 = Reserved
			0111 = C1OUT
RPC9	RPC9R	RPC9R<3:0>	1000 = Reserved
RPC2	RPC2R	RPC2R<3:0>	•
RPC4	RPC4R	RPC4R<3:0>	•
RPC3	RPC3R	RPC3R<3:0>	1111 = Reserved

Figure 9-13: Group 4 PPS output pin assignments

We now want to assign UART2 TX to RPB10 (RB10 pin). For output assignments, verify in data sheet table (shown partially in Figure 9-13) that a

PPS connection is possible. The table can be found in the PIC32MX250F128B data set under PPS Output assignments. For the library function PSS Output () arguments for this table are defined as group 4, and we are connecting RPB11 (or RB11) to U2TX for final function call of PSS Output (4, RB10, U2TX). This can be also be visualized with the following picture where an input multiplexer for RPB10 pin to U2TX is controlled. The function sets an SFR that contains the RPB10R bits with the value 0011, enabling the specific pin RPB11 pin to U2TX peripheral connection. See Figure 9-14.

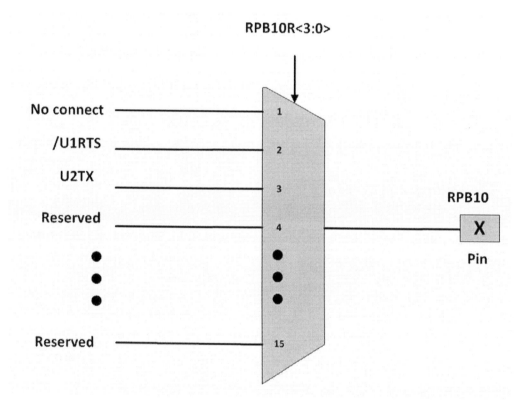

Figure 9-14: Remappable Output with RPB10

As a reminder the benefits of PPS are as follows:

- Allows optimal usage of on-chip peripherals
- Allows supported digital peripherals to remap to selected RP pin
- Pin assignment is set in software
    - Outputs assign a peripheral to a specific pin
    - Inputs assign a pin to a specific peripheral
    - Multiple input peripheral functions on one pin is supported

A UART Application Software Library

The experiments use a new application library #include "CONU2.H" in MAIN and CONU2.C driver source code in the project that allows use of the UART2 serial port from the user application code side. The CONU2 library functions used in the serial communications are:

- initU2 ( ) – initializes the serial port using UART2 of PIC32MX for RB11 and RB10 or pins 22 and 21 respectively of the Microstick II. Pin 22 is the input receive (RX) for the Microstick II UART2 and pin 21 is the output transmit (TX) for the Microstick II UART2. UART2 is configured for 2400 BAUD, 8 bit data, no parity and 1 stop bit or short hand 24008N1.

- putU2(character) – writes character out to the serial port

- putsU2(string)- writes an entire string out to serial port

- Clrscr() – this macro clears screen using a VT100 know escape sequence
- Home() – positions cursor to home position on screen using a known VT100 escape sequence

- Character getU2 ( ) – waits for a new character to arrive to the serial port and returns it.

We have been adding to our library in earlier chapters. To use the CONU2 library you must configure the PPS for UART2 in your main code. Our library can work will different system clock settings (this has be changed in System.h, Delay.h and CONU2.h). Our enhanced applications library now looks as follows with UART2 capabilities:

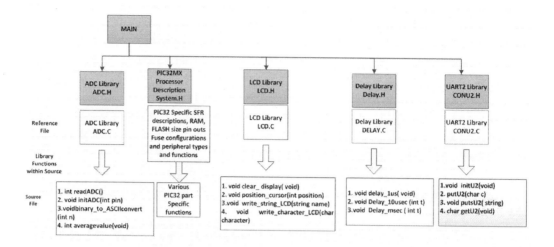

Figure 9-15: Application Library with UART functions shown

Experiment 1 "Implementing a RS-232 Serial interface with a PC

The first experiment will use the UART2 of the PIC32MX to communicate with a PC using RS-232 and the PC running HyperTerminal Software. The experiment uses the overall connection scheme of Figure 9-3. The application initially sends a two ACSII character strings to the PC "The Microstick UART2 Experiment" and "Type a character and watch return". After this any type character on the PC keyboard, is displayed these characters on the HyperTerminal Display window. The ASCII characters typed on the PC keyboard are sent via serial Communications to the Microstick II UART and upon reception are "echoed back" (or transmitted) to the PC for HyperTerminal display.

Hardware Configuration

The UART2 supports an RS-232 serial port with the proper physical interface. ACRONAME (http://www.acroname.com/) provides the ideal physical interface and is pretty inexpensive. The ACRONAME RS-232 interface is shown in Figure 9-3. Only four connections are required to work with the device. They are RX and TX from UART2 and +3.3V and GND. A schematic diagram for the Microstick II ACRONAME side of the serial interface is shown. (See Figure 9-16)

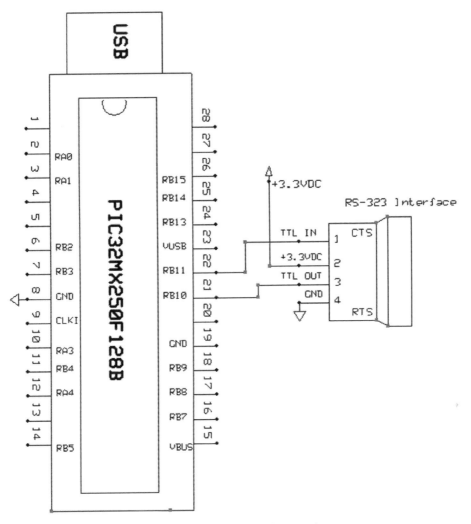

Figure 9-16: Experiment1 Schematic

The ACRONAME interface or "Brain stem" come in two formats RS-232 and USB. We will use both.

The ACRONAME 4 pin interface is on .100 "centers and is a female in line connector. A way to connect this to our prototype board is to use two 4 pin male headers soldered together at right angle where one side plugs into the Solderless breadboard and the other to the ACRONAME (see prototype Figure 9-17).

Figure 9-17: RS-232 Prototype

Assemble the hardware using both connection diagram (Figure 9-16) and the prototype construction shown in Figure 9-17. Note that an external power source of +3.3V DC is required for the ACRONAME. Be sure that all grounds between the ACRONAME, Microstick II (pin 8) and External Power are common.

Bringing up HyperTerminal

Bring up the HyperTerminal application. Configure HyperTerminal for 24008N1 with no hardware handshake. Use Figures 9-18 for new connection name and 9-19 for serial port configuration guidance.

Figure 9-18: Bringing up HyperTerminal

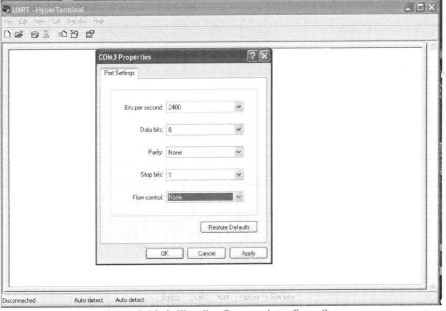

Figure 9-19: Setting the Com port configuration

Beginner's Guide to Programming the PIC32

Executing the code

Navigate to Chapter 9 folders and open up UART MPLAB X project experiment 1. Build and download into Microstick II. Make sure external +3.3V is supplied. Use the ACRONAME RS-232 electrical interface and an RS-323 cable to connect to PC serial port.

Once everything is connected and HyperTerminal is running, run the Microstick II UART program and you should see a screen introduction and then be able to type on the PC in the HyperTerminal window and see whatever you typed echoed back (see Figure 9-20).

Congratulations! You have a working serial port interface.

Figure 9-20: Running the Experiment

Figure 9-21 shows a snapshot of the UART2 initialization code that exists in CONU2.c. There are 3 UART SFR that need initialization; U2BRG for baud rate, U2MODE for mode, and U2STA for status. Refer to PIC32MX datasheet to understand specific SFR bit settings notations. Note that the U2BRG rate is set using formula (Peripheral Bus clock/Desired baud rate/16)-1.

```
#include <plib.h>
#include "conU2.h"
#define FPB 8000000
#define BAUDRATE 2400
#define BRGVAL ((FPB/BAUDRATE)/16)-1
#define U_ENABLE 0x8000        // enable the UART
#define U_TXRX   0x1400        // enable transmission and reception
// init the serial port (UART2, 2400, 8, N, 1 )
void initU2( void)
{
        U2BRG = BRGVAL; // BAUD Rate Setting for 9600
        U2MODE = U_ENABLE;
        U2STA  = U_TXRX;
        U2MODEbits.BRGH = 0; // standard mode
} // initCon
```

Figure 9-21: UART2 Initializations

Figure 9-22 shows the main loop. The home and clear functions are shown as well as the setup of the initial "splash screen". After that the code enters a continuous loop it simply gathers an incoming character from the PC and outputs this same character back.

```c
/****************************************************************/
#include "system.h"
#include "CONU2.h"
//****************************************************
static char      rxchar='0';              //received character
/********************************
         main entry point
********************************/
int main ( void )
{       /*      Initialize PPS */

PPSInput(2, U2RX, RPB11);        //Assign U2RX to pin RPB11
PPSOutput(4,RPB10, U2TX);        //Assign U2TX to pin RPB10

 ANSELA =0; //make sure analog is cleared
 ANSELB =0;
 initU2();
  rxchar =0;

// Hyperterminal or Tera Term Startup Inro Screen
        clrscr();  //clear hyper terminal screen
    home();
    putsU2("The MicroStick UART2 Experiment");
    putU2(0x0a);  //carriage return /line feed
    putU2(0x0d);
    putsU2("Type a character and watch the return");
    putU2(0x0a);  //carriage return /line feed
    putU2(0x0d);


        /*   endless loop*/
        while (1)
        {
        rxchar = getU2();
        putU2(rxchar);    }

}
```

Figure 9-22: Main Code

Using USB instead of RS-232

With the proper interface a UART can still be used for USB as well as RS-232. If you're PC does not include an RS-232 port –no worries. A USB (Universal Serial

Bus) port can easily be adapted to your serial communication needs. Again ACRONAME has a solution (S27), with an UART to USB interface. We will use the scheme shown in Figure 9-5.
Additional software configuration is needed. For USB, software INF files, and USB drivers will need to be installed on your PC--- ACRONAME, again, provides all software needed with their product.

Use the schematic to wire your prototype.

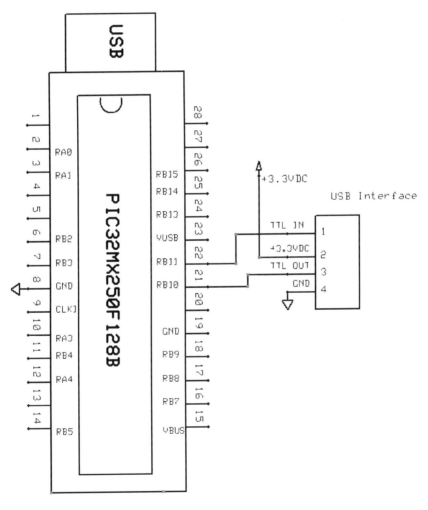

Figure 9-23: USB Exercise 1 Schematic

The final prototype should look as follows (see Figure 9-24)

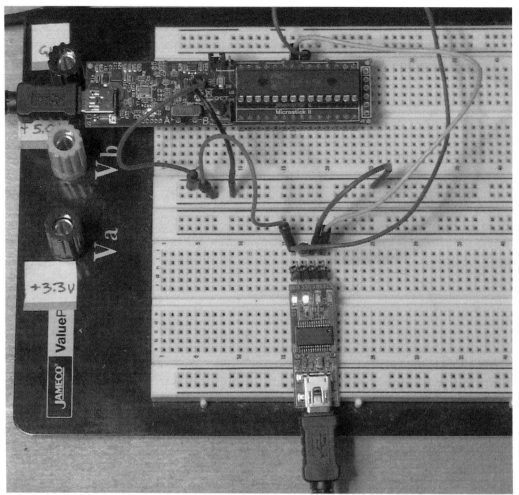

Figure 9-24: USB Prototypes

Bringing up Tera Term

Down load the Tera Term application. Unlike HyperTerminal it is open source and available for free from a number of internet sites. Configure Tera Term for 24008N1 with no hardware handshake. Use Figures 9-25 for new connection name and 9-26 for serial port configuration guidance. The configuration shown is COM1 2400N81.

When you first connect the USB-to-UART S27 ACRONAME interface, Windows will prompt you for a driver. Navigate to the ARCONAME supplied drivers that are supplied with the product and download these to your computer. Once installed it will essentially represent a new COM. Use this new com port and then use the serial port configuration procedure given earlier with this new COM port.

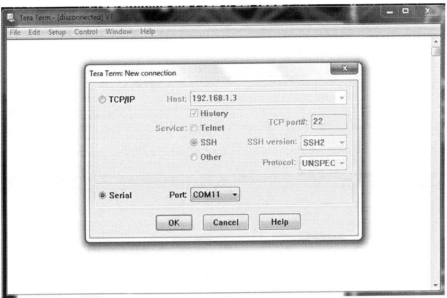

Figure 9-25: Bringing up Tera Term

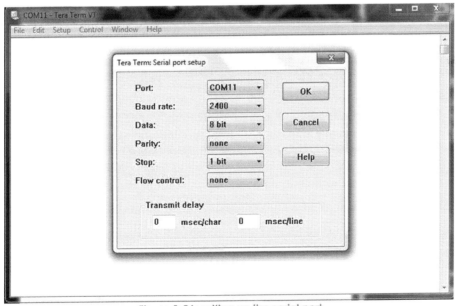

Figure 9-26: setting up the serial port

All operations from here on should be identical to the earlier experiment. There isn't even a need to change the Microstick II code from what was used earlier. See Figure 9-27 for results.

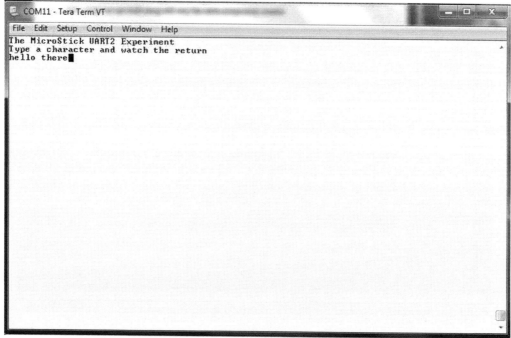

Figure 9-27: USB operation

Experiment 2 "Using Interrupt capability with UART2"

The first experiments with UART2 were done without interrupts. It was not so much a problem with UART data transmission, because the latency in transmitting characters is small. The problem comes up in reception. You just don't know when a character will appear from an outside communicator. Using the non-interrupt function getU2 (), the code will need to poll the UART2 continuously for a received character and remain there until one is received. This brings all other operations to a halt. The way around this is to use interrupts for UART2 receive.

Once a character is received the UART2 interrupt service is invoked where the character is retrieved from the UART and stored into a global variable rxchar and a flag is set got_rx. It is simple enough and provides a lot of flexibility for the microcontroller to go off and do other things instead of waiting on a character. The operations are identical to experiment 1. You can use either the USB or RS-232 setups.

Navigate to Chapter 9 folders and open up UART MPLAB X project experiment exercise 1 interrupt. Build and download into Microstick II. Make sure external +3.3V is supplied. Following the instructions covered earlier you should see similar responses.

The differences in code are in the main function. We use the interrupt service routine for UART2 with a priority level of 2 and define two global variables (rxchar and got_rx) to facilitate communication between the interrupt and main code. The ISR retrieves an 8 bit character from U2RXREG SFR and stores it in rxchar. It then sets got_rx flag as a semaphore to the main code that a character is received and available.

```
static char      rxchar='0';            //received character

unsigned char got_rx =0;                //rx char flag
//RX U2 interrupt
void __ISR(_UART_2_VECTOR, ipl2) UART2BHandler(void)
{


            rxchar =   (char) (U2RXREG & 0x00ff);


            got_rx =1;
            // Clear the UART RX interrupt flag bit
        IFS1bits.U2RXIF = 0;


}
```

Figure 9-28: UART2 Interrupt RX ISR

The main code is different in that during the final loop it tests the got_rx flag and if set, clears it and sends character to the PC as an echo response. Note that no matter what happens the Nop () function is always executed. This Nop () function is meant to represent other activities the microcontroller can be involved in. Note that just after UART2 initialization and before the main splash screen is written, the library function to enable the vectored interrupt control is used;

INTEnableSystemMultiVectoredInt()

The UART2 interrupt priority is set to 2, the interrupt for UART2 receive is enabled and the associated UART2 RX Interrupt flag is cleared, by writing directly to appropriate SFRs.

Beginner's Guide to Programming the PIC32

```c
int main ( void )
{       /*      Initialize PPS */

PPSInput(2, U2RX, RPB11);           //Assign U2RX to pin RPB11
PPSOutput(4,RPB10, U2TX);           //Assign U2TX to pin RPB10
 ANSELA =0; //make sure analog is cleared
 ANSELB =0;
 initU2();
rxchar =0;
got_rx= 0;

 INTEnableSystemMultiVectoredInt();
// interrupt setting
IEC1bits.U2RXIE =1;
IPC9bits.U2IP =2; //set priority to 2
IFS1bits.U2RXIF =0;

// Hyperterminal Startup Inro Screen
        clrscr();  //clear hyper terminal screen
    home();
    putsU2("The MicroStick UART2 Experiment");
    putU2(0x0a); //carriage return /line feed
    putU2(0x0d);
    putsU2("Type a character and watch the return");
    putU2(0x0a); //carriage return /line feed
    putU2(0x0d);

        /* endless loop*/
        while (1)
        {
    if (got_rx==1){
        got_rx =0;
        putU2(rxchar);
            }
        Nop(); //do something else
        }
}
```

Figure 9-29: Interrupt Code Main Function

Review of PIC32MX UART capability:

Questions:

1. Define UART

2. Define RS-232

3. Define DCE

4. Define DTE

5. What is baud?

6. What does 24008N1 mean?

7. What is the purpose of HyperTerminal or Tera Term?

8. How do we get the PIC which is a 0<-> +5V comply with RS-232 -12 to +12V standards?

9. What is the highest baud rate achievable by PIC? What clock is necessary?

10. What is the minimal amount of RS-232 pin connections to achieve two way communications?

11. What the purpose is of start and stop bits?

12. What is the purpose of parity?

13. How many UART does the PIC32MX have?

14. What is meant by half duplex?

15. What is meant by full duplex?

16. How is RS-485 different from RS-232?

17. What are the benefits of PPS?

18. How many reprogrammable pins does the PIC32MX250F128B have?

19. Where does PPS occur in code?

20. Define USB?

21. Can USB be used in place of legacy RS-232 communications? If so explain how.

22. Why use interrupts for UART?

Exercise:

1. Configure an UART1 to replace UART2 in both non-interrupt and interrupt exercises
2. Change baud rate to 1200

Chapter 10 – The Synchronous Peripheral Interface (SPI)

In this chapter we will develop prototype hardware and software to understand and exercise PIC32MX Synchronous Peripheral Interface or SPI. For the exercise we will use SPI2 to communicate (write and read data) using an EEPROM (Electrically Erasable Electrically Programmable ROM) that has an SPI interface. In addition we will add this ability to our growing application library.

The PIC32MX has two fully independent SPI peripherals: SPI1 and SPI2. In order to use any of these SPI you first must configure programmable PIC32MX pins using the Peripheral Pin Select or PPS. We learned about PPS in Chapter 9. If you need to review or to learn PPS or serial communications basics we suggest that you first complete Chapter 9 Serial Communications.

The SPI itself is a serial communication just like the UART. However it uses a shared clock between communicating entities making it a synchronous serial communications. In most cases the clock rates for SPI are higher than UART rates (up to 20 MHz) and since the SPI has no framing overhead (like start and stop bits) it's an ideal choice for efficient high point to pint speed transfers. A couple of examples are shown in Figure 10-1. SPI can also be half or full duplex allowing for transfers in one or both directions simultaneously.

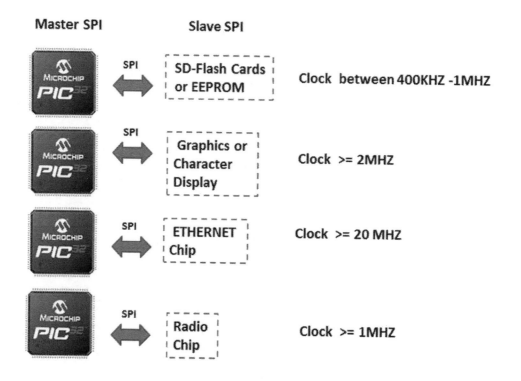

Figure 10-1: SPI Interface Examples

Let's take a closer look at the SPI interface. It requires four signals: SDO (serial data out), SDI (serial data in), SCLK serial clock, and /CS (Chip Select). The "Master" side, of an SPI communication initiates the communications and is the sole source of Clock and CS. The "Slave" communication responds to the Master (see Figure 10-2). There is no protocol only data movement. Both sides function as shift registers where the common clock is used to shift data from one transmitting register (SPI BUF) into the receiving register (SPI BUF). The CS signal is used to enable slave peripheral for communications and can be duplicated to allow the SPI communication interface to be shared between one master and many slaves (however, only one slave at a time). In summary:

- Simple interface
- Uses four signals
    - /CS, SDO, SDI, CLK
- Based upon a shift register concept
- Very high speed (up to 20MHZ with PIC32MX)

- Robust and very noise immune

- The Master SPI could support multiple slaves SPI by using a separate CS for each slave and then communicating with them one at a time.

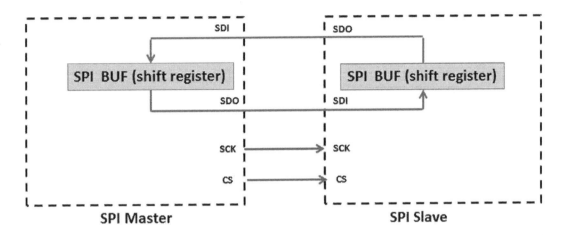

Figure 10-2: SPI Serial Communications

Introducing a Serial EEPROM

Our experiment will use a serial EEPROM to demonstrate SPI operation. Specifically we will use a Microchip 25LC256 EEPROM device that supports 256kbits or 32Kx8 bytes of nonvolatile memory. The EEPROM is a memory technology that allows us to store data under program control and retain the contents even when power is turned off. This can be real useful for user applications requiring password setting, storing calibration data, or storing data measurements as in a logging function. The 25LC256 uses an application library SPI interface to talk to the Microstick II PIC32MX Microcontroller. The API enables reuse in creating your own EEPROM applications.

For this library each 25LC256 storage word is 32 bits or 4 bytes. Given the byte size of the 25LC256 this means we can store up to 8K 32 bit words total. The quad byte storage and retrieval makes sense since the PIC32MX is a 32 bit machine. In general, for your experiments, you need to establish what data you want to store and how that data is organized over 8K 32 bit words. Figure 10-3 shows the physical EEPROM. It is an 8 pin DIP. You already can recognize the basic SPI signals /CS, SCK, SI and SO. In addition to this there are some more control signals. There is a write protect (/WP) control line to physically lock out the Microcontroller from writing to the EEPROM. /HOLD suspends a command/data sequence, and puts the device in a suspended state that does not require restarting the sequence once /HOLD is released. In our experiment we will just use the basic signal set all others will be tied high or disabled.

- $\overline{CS}$ – Chip Select
- SCK – Serial Clock
- SI – Serial Input
- SO – Serial Output
- $\overline{WP}$ – Write Protect
- $\overline{HOLD}$ – Hold Enable

Figure 10-3: SPI EEPROM

Experiment 1- Hardware

The PIC32MX allows for user programmable RP pins to be configured with I/O of select internal digital peripherals. This is done through the PPS function covered in Chapter 9. We will configure RP pins 2, 13 to function as the interface for SPI2, namely, SO and SI. RB15 is already configured for CLK, so no PPS is necessary. RB 15 is configured as a normal digital output for /CS. /HOLD and /WP are not required and are wired high to disable on the EEPROM device. The final hardware schematic is shown in Figure 10-4.

Figure 10-5, 10-6 and 10-7 show the 25LC256 EEPROM instruction set and a read/write byte sequence using SPI. Note that all addresses to the EEPROM are sixteen bit. Upon /CS becoming active the microcontroller first sends the specific command (see Figure 10-5) to the EEPROM, followed by address for 32 bit data transfer. What happens next depends on whether the command is a read or write operation; if read the EEPROM will output data to Master to read, if it is a write operation the Master will output data from the EEPROM to store. Most EEPROM are not typically high speed devices and a memory write cycle may take several milliseconds. The API handles this. The API requires that all usable addresses are even addressed and occur only on 4 byte boundaries.

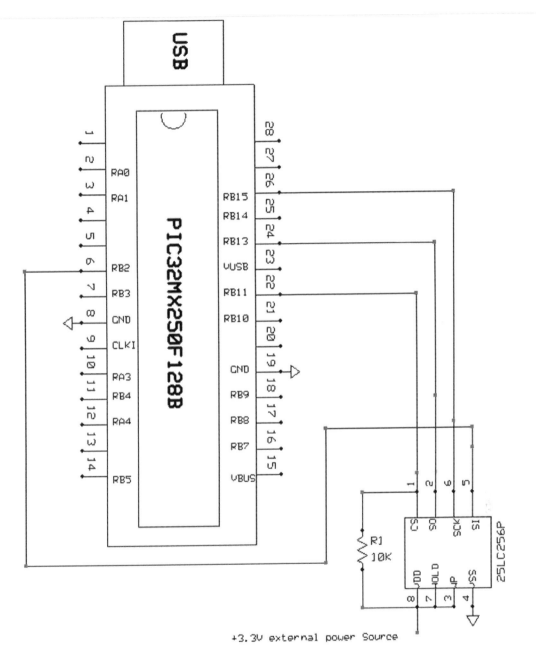

Figure 10-4: Schematic

Instruction Name	Instruction Format	Description
READ	0000 0011	Read data from memory array beginning at selected address
WRITE	0000 0010	Write data to memory array beginning at selected address
WRDI	0000 0100	Reset the write enable latch (disable write operations)
WREN	0000 0110	Set the write enable latch (enable write operations)
RDSR	0000 0101	Read STATUS register
WRSR	0000 0001	Write STATUS register

Figure 10-5: Microchip 25LC256 EEPROM Instruction Set

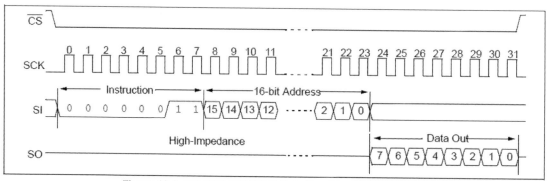

Figure 10-6: 25LC256 EEPROM Byte Read Sequence

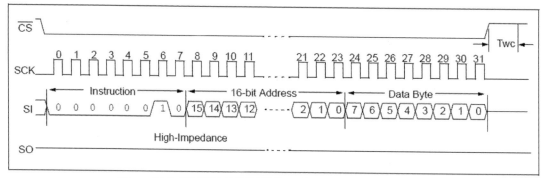

Figure 10-7: 25LC256 EEPROM Byte Write sequence

Experiment 1- The Software an EEPROM Library

Ok, the hardware has been discussed, let's now discuss the software. The EEPROM library is fairly straightforward, and uses 32 bits of data as its argument (4 byte read or 4 byte write). Take a quick look at the functions listed below.

Using the EEPROM library function we can do the basic initialization of SPI2 to access the EEPROM and then read and write from it once it is initialized.

- **InitSEE ()** - initializes the EEPROM for access using SPI2. This must be called first before any other EEPROM library function is called. NVM stands for "Non-Volatile Memory".

- **int readSEE (address)-** SPI2 reads EEPROM at address and returns the 32 bit value stored at that location

- **WriteSEE (address, data)** – SPI2 writes 32 bit data value to EEPROM at the designated address.

We have been adding to our library in other earlier chapters. To use the SEE.h library components you must configure the PPS for SPI2 in your main code. Our library can work will different system clock settings (this has be changed in System.h, Delay.h and CONU2.h, SEE.h). Our enhanced applications library now looks as follows with NVM capabilities: Our total application library has now grown. A snapshot is provided.

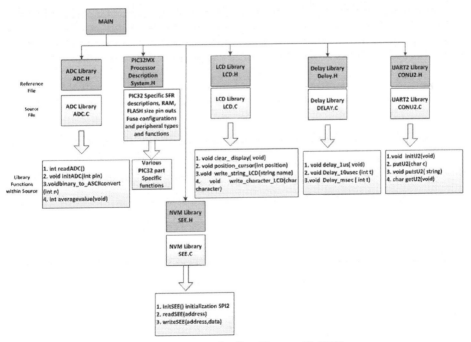

Figure 10-8: Application Library with NVM

Beginner's Guide to Programming the PIC32   Page 201

An example project in MPLAB X chapter 10 uses NVM for our experiment. It configures SPI2 interface SDI2 and SDO2 using PPS. It the initializes the SPI2 Special Function Registers SPI2CON(control) and SPI2BRG (baud rate) for correct operation, finally it writes a 32 bit value to an even address in EEPROM (located on a 4 byte boundary) and then reads this value back from the EEPROM and compares values. A flowchart is provided.

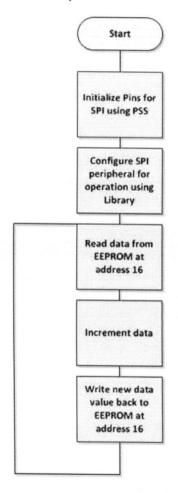

Figure 10-9: Flowchart

Experiment 1- How it Works

The application uses a primary main source file SEETEST.c and library files System.h, SEE.H and SEE.C. Let's first examine SEETEST.c and how it programs the PPS, and then let's examine SEE.C for how the SPI2 is initialized. Finally, let's examine the overall main program operation.

The main.c starts with setting all pins from default analog to digital using ANSELA and ANSELB =0, it then performs PPS (using PPSInput () and PPSOutput () Microchip Peripheral Library functions) to program two PIC32MX programmable pins for configuring SPI2 data out, SPI2 data in. The PIC32MX250F128B has SCK2 already assigned as RB15. No PPS is necessary; we simply make sure that SCK2 is assigned as an output because it will generate a clock for the EEPROM SPI interface. / CS is programmed in SEE.C using digital output pin RB11 (See Figure 10-13).

```
ANSELA =0;  //make sure analog is cleared
ANSELB =0;
PPSInput( 3,SDI2,RPB13);      //Assign SDI2 to RBP13
PPSOutput(3, RPB2, SDO2);     //Assign SDO2 to pin RPB2
```
Figure 10-11: Programing the PPS for SPI2

The code then uses the application library function initSEE () to configure the SPI2 peripheral, and then readSEE(address) to read and writeSEE (address, data) to write 32 bit data from/to the EEPROM (see Figure 10-12). In this case the initial value of data is retrieved from an even addresses of EEPROM, address 32, and then is incremented and written back to same EEPROM address. The value is stored, then read back and stored into global variable data.

```c
#define SCK2    _TRISB15    // tris control for SCK2 pin26
//note CSEE define and set in SEE.C as RBP11
    int data;
    SCK2 =0;        //set as output
    // initialize the SPI2 port and CS to access the 25LC256
    initSEE();
    // main loop
    while ( 1)
    {

        // read current content of memory location
        data = readSEE( 32);


        // increment current value
        data++;           // <-set breakpoint here

        // write back the new value
        writeSEE( 32, data);
        //address++;

    } // main loop
} //main
```

Figure 10-12: 25LC256 EEPROM initialization read and write using SPI2

The library function InitSEE () initializes SP2 through its Special function registers SPI2CON (control) and SPI2BRG (baud rate). This code is shown in SEE.C (see Figure 10-13).

```c
// I/O definitions
#define CSEE      _RB11       // select line for Serial EEPROM
#define TCSEE     _TRISB11    // tris control for CSEE pin

// peripheral configurations
#define SPI_CONF  0x8120      // SPI on, 8-bit master,CKE=1,CKP=0
#define SPI_BAUD  15          // clock divider Fpb/(2 * (15+1)) 250KHZ

// 25LC256 Serial EEPROM commands
#define SEE_WRSR   1          // write status register
#define SEE_WRITE  2          // write command
#define SEE_READ   3          // read command
#define SEE_WDI    4          // write disable
#define SEE_STAT   5          // read status register
#define SEE_WEN    6          // write enable

void initSEE( void)
{
    // init the SPI2 peripheral
    CSEE = 1;                 // de-select the Serial EEPROM
    TCSEE = 0;                // make SSEE pin output
    SPI2CON = SPI_CONF;       // enable the peripheral
    SPI2BRG = SPI_BAUD;       // select clock speed
}// initSEE
```

Figure 10-13: CS and SPI2 Initialization

The contents and setting of SPI2BRG and SPI2CON are shown in Figures 10-13. SPI2CON is set to 0x8120, SPI on, 8 bit master, CKE (clock enable) =1, and CKP (clock polarity) =0. SPI2BRG is set to 15 representing a clock rate of about 250 KHz at a peripheral Bus rate of 8MHz. These settings and their meanings are detailed in PIC32MX250F128B datasheet.

Experiment 1 Execution

Build the prototype using schematic in Figure 10-4. A 10 k resistor and a 25LC256 8 pin DIP EEPROM are needed. In addition an external +3.3V DC power supply is required to power up the EEPROM (remember that the Microstick II does not supply external power). Make sure the ground for the EEPROM is common to the ground for Microstick II. An example prototype assembly is shown in Figure 10-14. Connect the Microstick II to the PC's USB port.

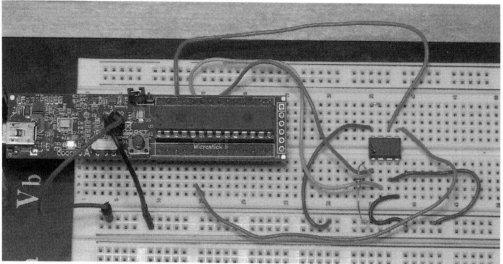
Figure 10-14: Assembled Prototypes

Open the folder for Chapter 10 MPLAB X project NVM. Build and download the code into the Microstick II. A snapshot of the main code is shown. Set break point and designated location of operation data++.

```
while ( 1)
{
    // read current content of memory location
    data = readSEE( 16);

    // increment current value
    data++;        // <-set breakpoint here

    // write back the new value
    writeSEE( 16, data);
    //address++;

} // main loop
```

Figure 10-15: Main Code

Set a watch for data using the debugger. Set breakpoints designated. Reset the device and then run. You should break at the successful compare point. Double check the data value under watch. If you don't break you have a wiring error. Please go back and double check your prototype using the schematic.

If successful you now have a general API and hardware configuration for the PIC32MX and 25LC256 EEPROM. This setup can be reapplied for use in a number of applications.

Review of SPI Key features:
- Supports four wires SDO, SDI, SCLK, /CS
- Data can move in both directions simultaneously.
- Master supplies Clock and initiates data transfer
- Simple synchronous interface used extensively in microcontroller peripheral communications
- Capable of high speed transfers (clock rate dependent)
- SPI Peripheral functions as a digital shift register

Review of EEPROM Key Features:
- EEPROM retains memory content when power is off
- Use in setting passwords, access code, and calibration data

Review of PIC32MX SPI Key Features:
- PIC32MX has two available SPI peripheral SPI1 and SPI2
- PPS is used to configure programmable pins for SPI interface
- SPI2 is programmed through the bit setting in two SFR registers SPI2CON1 and SPI2STAT
- An API library is provided for SPI2 initialization and reading and writing 4K 32 bit values to the 25LC256 32Kx8 EEPROM. Addresses must be on an even 4 byte boundary

Exercise:
1) Try different values and different address for NVM project
2) Describe an application that can use an EEPROM
3) Configure NVM project to work with SPI1 versus SPI2
    a. Change PPS for SPI1 pins
    b. Rewrite Library for operation with SPI1

Chapter 11 – Using PWM for Tone Generation

In this chapter we will develop prototype hardware and software to understand and exercise the Pulse Width Modulation (PWM) associated with the Microstick II PPIC32MX Output Compare Module (OC). Pulse width modulation (PWM) is a powerful technique for controlling analog circuits with a processor's digital outputs. PWM is employed in a wide variety of applications, ranging from measurement and communications to power control and conversion. Tone generation for ring tones or simple sound alerts is an important part of consumer electronics

Because PWM operations are important for a large variety of microcontroller applications, Microchip's PIC32MX provides us with an impressive arsenal of internal pulse generation peripherals: five Output Compare Modules. The Output Capture Module (OC) can generate continuous frequency pulse waveforms with changeable duty cycle on output. Duty cycle refers to the amount of time the pulse is high over the period of the pulse frequency. For instance a 50% duty cycle is the pulse on for half the period, while a 100% is having the pulse on for the whole period. Minimal software is required to operate this module set, as a lot of what is done is achieved mostly in hardware. We will cover the use of this module in with two demos; one that that illustrates LED brightness control and the other tone generation.

Some PWM applications are:

- Speed control for DC motors (PWM duty cycle controls motor average DC level or speed).

- Audio generation where pulse width represents the current DC amplitude of the audio signal and a low pass filter is used to remove the pulse frequency.

- Tones generator with a buzzer. PWM duty cycle is fixed but pulse frequency (changes for different tones).

- Light dimming (PWM duty cycle controls average DC current through LED thereby setting its brightness).

Output Compare Module (OCx)

The Microstick II Microcontroller PIC32MX250F128B has five Output Compare modules OCx (OCx, where X =1, 2, 3, 4, 5). Any OCx can be mapped to a number of the PPS pins. For basic familiarity with PPS please review Chapter 9. A block diagram of an OCx is shown (Figure 11-1).

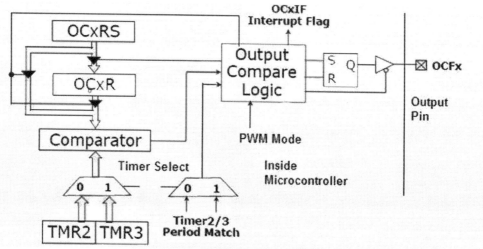

Figure 11-1: Output Compare Peripheral

The OCx module can use either the PIC32MX250F128B 16 bit Timer 2 (TMR2), or Timer3 (TMR3) as a time base, and the period setting for the output pulse waveform. PR2 is the 16 bit period setting for Timer 2 and PR3 is the 16 bit period setting for Timer 3. In our first example we will use Timer 2. In our second example we will use the combined 32 bit Timer2/3. Timers were covered in Chapter 6. The OCxRS and OCxR registers are loaded with a 16 bit or 32 bit value to control the width of the pulse generated during the output period. This value is compared against the Timer during each period cycle. The OCx output starts high and then when a match occurs OCx logic will generate a low on output. This will be repeated on a cycle by cycle basis (See Figure 11-2).

This is OCx Pulse Width Modulation (PWM) mode. An important notion in thinking about PWM is that it allows dynamic changing of the DC voltage level to a load. Wait a minute you say --- where's the DC voltage here? Think of the DC voltage as average value voltage value of the pulsed DC across the waveform period. For example, if we have pulse high for 50% of the period consistently, and the pulses are +3.3V then we are supplying on the average 50% of 3.3V to the load or 1.65V. In order to change the pulse width while the system is running we simply write into OCxRS and it automatically loads into OCxR on the next cycle.

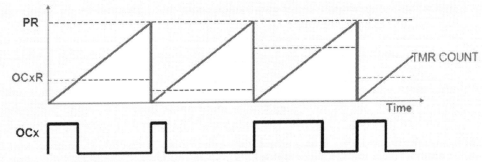

Figure 11-2: Output Compare Pulse Generation

As discussed there are many ways that PWM can be used. We will first look at how to apply this idea in demonstrating LED Brightness control using Pulse Width Modulation. The PWM controls the DC level to an LED connected to the output pin of OC1 thereby changing its brightness (see Figure 11-3). A manual brightness control can be achieved by reading the digitized setting of a pot that is connected to an analog input of the PIC32MX. Its input analog voltage is digitized and then this digital value is used to set the PWM duty cycle, thereby changing the PWM output at the OC1 pin.

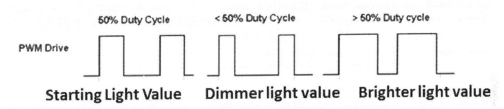

Figure 11-3: Example PWM Waveforms for lighting control

Exercise 1 LED PWM Experiment

The Experiment uses OC1 module to PWM drive an LED. The code is contained in MPLAB X project PWM in file folder for chapter 11. We will use Microchip library peripheral functions to set PPS for connecting OC1 output to RPB7, set Timer2 for the PWM waveform period, turn on the OC1 using Timer2, and then set the PWM duty cycle. The specific peripheral library functions used in the main code are:

- **PPSOutput (1, RPB7, OC1)** - Connects the OC1 output to pin RPB7. PPS functionality was covered in Chapter 9.
- **OpenTimer2 (T2_ON |T2_SOURCE_INT | T2_PS_1_256, 0x400)** – This library function turns on Timer2 using the internal peripheral bus clock as source, sets prescaler to 256 and initializes the Timer2 period to 0x400. This period was chosen because it is exactly equal to 1024 or full count of the 10 bit ADC.

- **OpenOC1(OC_ON | OC_TIMER_MODE16| OC_TIMER2_SRC | OC_PWM_FAULT_PIN_DISABLE, 0x400, 0x400)** – This library function turns OC1 using Timer2 in 16 bit mode as clock, disabling PWM fault (this is only needed for certain motor controls), and setting its timer comparisons to 0x400. The period of Timer2 is preset to 0x400 the max value of the 10 bit ADC. A setting is any pot value measured by the ADC between 0-0x400 to adjust the PWM duty cycle. If setting is 0 then duty cycle is zero, if setting is 0x200 then duty cycle is 50%, if setting is 0x400 then duty cycle is 100%. Our initialization insures that an initial PWM of 50% to start with.

- **SetDCOC1PWM (value)** - This performs the adjustment function of setting PWM for OC1 to the int value.

There are also several ADC functions that we will use. These functions were introduced in Chapter 5 and are in the ADC library. If you need to, please review Chapter 5 to gain an understanding of ADC.

The actual library functions used in this example for ADC are as follows

- **Void initADC (int pin)** –this function initializes the ADC for 12 bit manual conversion using the "ANpin" as the single input pin. Pin possible settings are AN0, AN1, AN2, AN3, AN4, AN5, AN6, AN9, AN10, AN11, and AN12. Note AN4 and AN5 have dual purpose as debugger controls for ICSP. Only one channel can be initialized at a time.

- **Int readADC ()** –this functions forces a manual ADC conversion on the designated ANpin.

In addition the following library functions for delay are used. Delay was first covered in Chapter 6

- **Delaymsec(t)** - **t =1000 millisecond used for one second updates to OC1**

The hook-up schematic is shown in Figure 11-4. A 10K pot, LED, 1K resistor and external +3.3V supply is needed for the other side of the 10K. Assure that all grounds between Microstick II and external components are common. Build up the prototype per the schematic and the prototype is shown in Figure 11-5 to assist you.

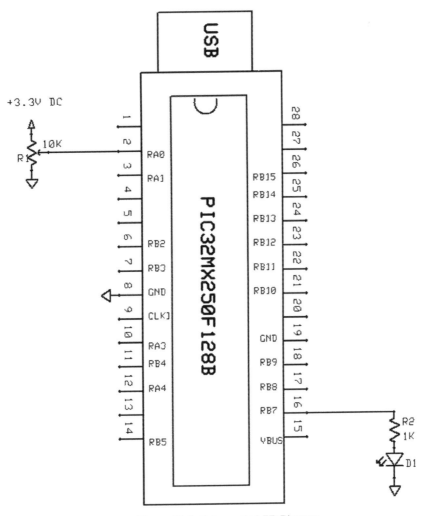

Figure 11-4: Schematic LED Dimmer

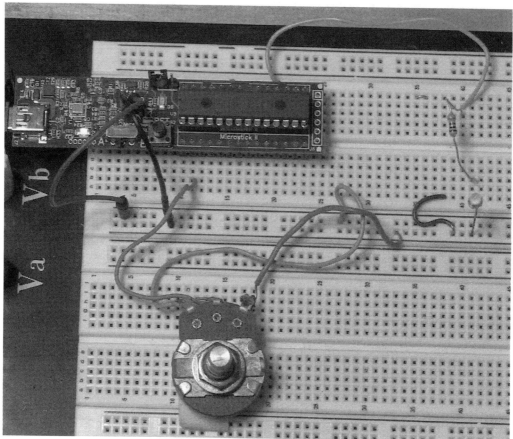
Figure 11-5: Picture of prototype

Software operation

We already discussed the Delay and ADC libraries and the OC1 configuration we are attempting to do. PPS was used to configure RPB7 to OC1. Let's put it all together using a flowchart.

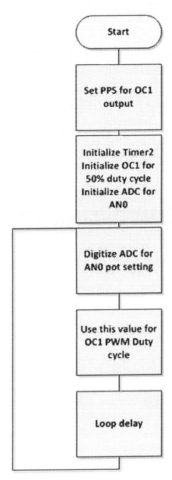

Figure 11-6: Flowchart

The PPS code that sets PIC32MX RPB7 pin 16 to be OC1 output. Group 1 is required based upon PIC32MX data sheet on PPS (again refer to Chapter 9 for more discussion on PPS).

PPSOutput (1, RPB7, OC1);

The main code uses two application libraries delay and ADC. System.h is also used to reference Microchip peripheral libraries, and set PIC32MX250F128B part specific configuration.

```c
#include "System.h"    /* generic header for PIC32MX family */
#include "adc.h"
#include "delay.h"
int adcvalue =0;
//note only valid pins are 2,3,4,5,6,7,24,25,26-> PIC32MX250F128B
//note 4,5 are used for ICSP, 2 is USer LED is J3 is used
int pin =2; //an0 pin 2


int main ( void )
{ /* ADC example- need potentiometer between +3.3V and
   *  Ground with wiper tied to pin 2 of PIC32MX         */
    ANSELA =0; //make sure analog is cleared
    ANSELB =0;
    PPSOutput(1,RPB7, OC1);
    OpenTimer2 ( T2_ON |T2_SOURCE_INT | T2_PS_1_256 , 0x400); //set period
    OpenOC1( OC_ON | OC_TIMER_MODE16| OC_TIMER2_SRC | OC_PWM_FAULT_PIN_DISABLE, 0x400, 0x400);
    SetDCOC1PWM(0x200); //inital 50% duty cycle
    //TRISBbits.TRISB7 =0; //Set bit 7 to ouput


    initADC(pin); //Initialize ADC for AN0 pin 2 of PIC32MX250F128B
    while (1)      /*  endless loop vary pot and capture reading*/
    {
        adcvalue =readADC (); //force a conversaion
        SetDCOC1PWM(adcvalue); //set PWM dutycycle to ADC value
        Delayms(1000);

    }

}
```

Figure 11-7: PPS, Timer2, OC1 and ADC Initialization and main Loop

Running the Light Dimmer Experiment

We have covered both the hardware and software so you are ready to run the prototype. Connect to the USB port of the PC. Connect your external +3.3V supply for use with the pot. Make the pot and LED circuit grounds common with the Microstick II ground. Navigate to Chapter 11 folders and use MPLAB X to open project PWM. Make sure Microstick II prototype is connected. Run debug project.

With the program running, as you turn the pot either clockwise or counter clockwise you should see the LED change brightness. If this does not occur please refer to schematic and prototype picture as you have a wiring error.

Exercise 2 Tone Generation

Let's now explore tones generation using OC1 driving a buzzer. In this situation the PWM duty cycle is fixed but pulse frequency (changes for different tones). This exercise will use a simple buzzer. The schematic is shown in Figure 11-12. Our software will configure the OC1 using 32 bit Timer operation to precisely configure the period of the required tone frequency. The PWM duty cycle will be fixed to 50%. The 'C' library function we use accepts tone frequency and tone duration as input arguments. By reusing this function we can build up larger ring tones. In fact several examples will be covered here; DO-RE-MI, TWINKLE-TWINKLE LTTLE STAR, and an assortment of ACTION Tones (alarm, hyperspace, robot).

Software Overview

The key sound generation function code is shown in Figure 11-8. It incorporates a number of Microchip Peripheral Library functions introduced in earlier exercises.

```
void Sound ( long frequency, int duration) {
    unsigned int period;
    period = (unsigned int) (SYS_CLK/frequency);
    OpenTimer23(T2_ON | T2_32BIT_MODE_ON | T2_PS_1_2,period); //32 bit time
    OpenOC1( OC_ON | OC_TIMER_MODE32 | OC_TIMER2_SRC | OC_PWM_FAULT_PIN_DISABLE , period,period );
    SetDCOC1PWM(period/2); //> 50% duty cycle
    Delayms(duration);
    CloseOC1();
}
```

Figure 11-8: Sound Generation Function.

The code takes frequency argument and generates the required 32 bit timer count using SYS_CLK/frequency. This count or period is used to set Timer2/3 (32 bit configuration) period. It then configures OC1 using the 32 bit timer, and sets PWM to 50%. The OC1 then generates the frequency tone for the required duration and before shutting down using CloseOC1 () function.

We use the Delayms (duration) function from the Delay API library. In order to achieve the required frequency resolution we modified System.h to use 8MHz RC oscillator with PLL (phase lock loop) to increase SYS_CLK to 18 MHZ. A code snippet from System.h is shown in Figure 11-9 that causes this configuration. Here FRCPLL is selected; we take the 8MHz clock, divide it by 4 to get 2 MHz, and then multiply by 18 for 36 MHz, and then divide by 2 to get an 18MHz clock.

```
#pragma config FPLLIDIV = DIV_4      // PLL Input Divider (4x Divider)
#pragma config FPLLMUL = MUL_18      // PLL Multiplier (18x Multiplier)
#pragma config FPLLODIV = DIV_2      // System PLL Output Clock Divider/2
// DEVCFG1
#pragma config FNOSC = FRCPLL        // Oscillator Selection Bits (Fast RC Osc with PLL)
```

Figure 11-9: System clock changes

Each of the tones we are trying to create is just a sequence of calls to the sound (frequency, duration) function as shown for DO-RE_MI and TWINKLE-TWINKLE LITTLE STAR in figure 11-10.

```
void doremi(void) {

            Sound(262,2000) ; //C6
            Sound(294,2000) ; //D6
            Sound(330,2000) ; //E6
            Sound(349,2000) ; //F6
            Sound(392,2000) ; //G6
            Sound(440,2000) ; //A6
            Sound(494,2000) ; //B6
            Sound(523,2000) ; //C7

}
void twinkletwinkle(void) {
            Sound(1047,500) ; //
            Sound(2093,500) ; //
            Sound(2093,500) ; //
            Sound(3136,500) ; //
            Sound(3136,500) ; //
            Sound(3520,500) ; //
            Sound(3520,500) ; //
            Sound(3136,1000); //

}
```

Figure 11-10: Ring Tones

It all comes together with Main.c. Note that PPS Output is configured as before for RPB7 pin 16 connection to OC1.

```c
int main ( void )
{
    ANSELA =0;  //make sure analog is cleared
    ANSELB =0;
    PPSOutput(1,RPB7, OC1);
     doremi ();
    twinkletwinkle();
   actiontones ();
while(1);



}
```
Figure 11-11: Main Code

Running the Sound Experiment

Build the prototype using the schematic shown in Figure 11-13. It should look similar to the project shown in Figure 11-12.

Figure 11-12: Sound Prototypes using AC Buzzer RDI-DTM-1206

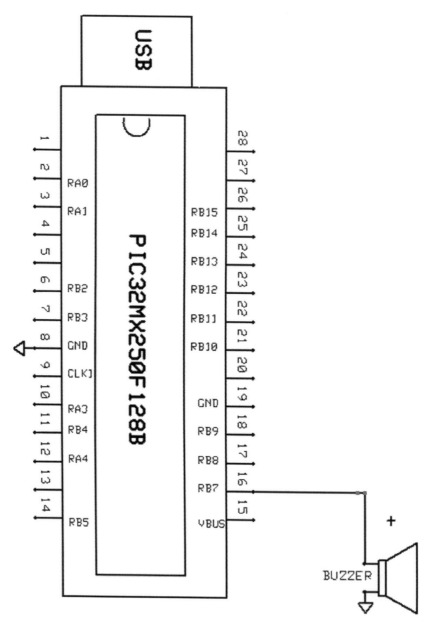

Figure 11-13: Tone Generation Schematic

The prototype should look like that shown in Figure 11-13. The buzzer is an AC buzzer, part number RDI-DTM-1206, specifically used for PWM. It is available from a number of electronic vendors. Note that the buzzer is polarized and the + side should be connected to Microstick II.

Navigate to Chapter 11 folder Sound and open PWM project with MPLAB X. Connect the Microstick II to PC, and run debug project. Enjoy the sounds!

Review of OC PWM Key features:

- There are 5 OCX peripherals within the PC32MX
- Access to the OC output requires a PPS setting.
- Once set for PWM mode the OC continuously outputs a digital pulse that is periodic with the duty cycle setting.
- Duty cycle setting is done via writes to OC1R and OC1RS
- The OC1 can use either Timer2 or Timer3 period for basic frequency and duty cycle timing

Exercises:

1) What does PWM stand for?
2) How is it used?
3) Run experiment 1 by turning pot to both extreme. Capture those using watch. Read and record values. What are they? Where do they originate form?
4) Describe an application that can use PWM
5) Describe how the OC is configured for sound
6) Try to write your own tones for exercise 2
7) Configure exercise 1 to work with OC2 versus OC1
   a) Change PPS for OC2 pins
   b) Rewrite Library for operation with OC2 SFR

Chapter 12 –RTCC (Real Time Clock Calendar)

In this chapter we will work with the RTCC (Real Time Clock Calendar) peripheral on the PIC32MX. We will explore how to set data, time, alarm, and use interrupt with the RTCC, and then integrate it with an LCD Display (this LCD was introduced in Chapter 5). For these exercises we will need to use the PIC32MX250F128B secondary oscillator input SOSCI (pin 11) and SOSCO (pin 12) with external oscillator components, a 32 KHz clock crystal with 33pf capacitors –more on this later.

The Real Time Clock Calendar

The Real-Time Clock and Calendar (RTCC) is intended for applications where accurate time must be maintained for extended periods of time with minimum to no intervention from the CPU. The peripheral is optimized for low-power usage in order to provide extended battery lifetime while keeping track of time.
It features a 100-year clock and calendar with automatic leap year detection. The range of the clock is from 00:00:00 (midnight) on January 1, 2000 to 23:59:59 on December 31, 2099. The hours are available in 24-hour (military time) format. The clock provides a granularity of one second.

The RTCC internal register interface with the PIC32MX Microcontroller is implemented using the Binary Coded Decimal (BCD) format. This simplifies the software, when reading and writing to the module, as each of the digit values is contained within its own 4-bit value.

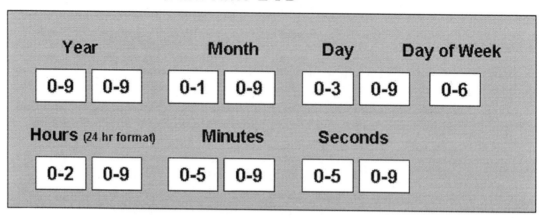

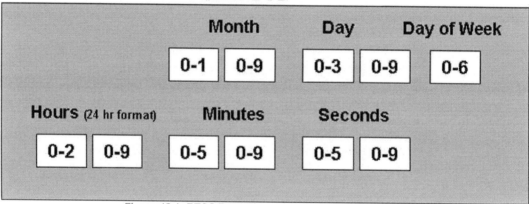

Figure 12-1: RTCC Date, Time, and Alarm Registers

The RTCC module is clocked by an external Real-Time Clock crystal that oscillates at 32.768 kHz. The clock crystal connects to the PIC32MX and works in conjunction with internal logic counters within the RTTC module to derive time (see Figure 12-2).

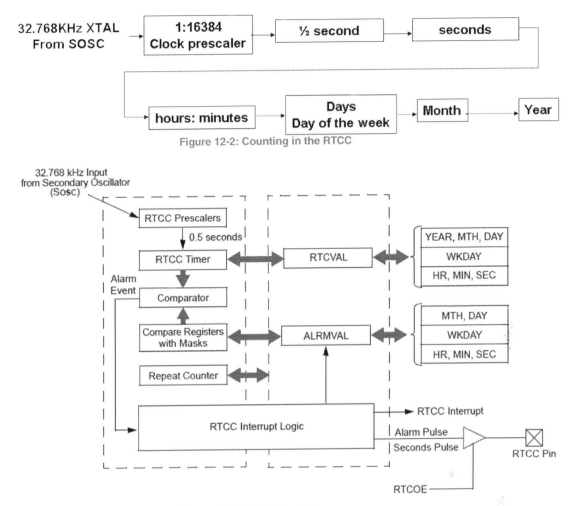

Figure 12-2: Counting in the RTCC

Figure 12-3: RTCC Block Diagram

The entire block diagram is shown in Figure 12.2. Time is incremented using the prescaler and 32 KHz oscillator to derive seconds, minutes, hours, days, day of the week, month then year and then stored in the internal register via RTCVAL pointer. The count is continuously compared against the alarm value of second, hour, minute, day, day of week, month using the ALRMVAL pointer. Note that year is not used in alarm. If alarm condition occurs an interrupt will notify the PIC32MX. Alarm time Intervals are configurable; you can test for the alarm on very Half Second, One Second, 10 Seconds, One Minute, 10 Minutes, One Hour, One Day, One Week, One Month or One Year. The alarm is cleared in software (usually as part of the RTCC ISR).

The RTTC has its own dedicated pin on the PIC3MX250F128B, this is pin 7 RB3. If enabled it can serve as a half second rate pulse stream to indicate clock is active or under alarm condition it can serve as "chime" pulse output. An Alarm condition can be made to repeat if user so desires.

The RTCC summary feature set includes:

- Time: Hours, Minutes and Seconds
- 24-Hour Format (Military Time)
- Visibility of One-Half-Second Period
- Provides Calendar: Weekday, Date, Month and Year
- Alarm Intervals are configurable for Half a Second, One Second, 10 Seconds, One Minute, 10 Minutes, One Hour, One Day, One Week, One Month and One Year
- Year Range: 2000 to 2099
- Leap Year Correction
- BCD Format for Smaller Firmware Overhead
- Optimized for Long-Term Battery Operation
- Alarm repeat with decrementing counter
- Alarm with indefinite repeat
- Requirements: External 32.768 kHz Clock Crystal
- Alarm Pulse or Seconds Clock Output on RTCC pin

Configuring and using the RTCC

We will use the Microchip peripheral library routines to initiate clock operations, set time/date, read current time/date. This library allows for a comprehensive C Union/Structure to represent time and date in a number of formats: BCD, byte, short word (16 bits) and long word (32 bits). This feature is very convenient in setting and reading RTCC date and time contents. Let review these structures first since they are heavily used in the RTCC library.

- **RtccDate** –this union/structure for read/write of date for the RTCC. Its membership includes the following. Any individual member is directly accessible.

    - unsigned char wday - BCD codification for day of the week, 00(SUN)-06 (SAT)
    - unsigned char mday - BCD codification for day of the month, 01-31
    - unsigned char mon - BCD codification for month, 01-12
    - unsigned char year - BCD codification for years, 00-99
    - unsigned char b[4] - byte access
    - unsigned short w[2] - 16 bits access

- o unsigned long l - 32 bits access

- **RtccTime** –this union/structure for read/write of time for the RTCC. Its membership includes the following. Any individual member is directly accessible.

    - o unsigned char rsvd -reserved for future use, should be 0
    - o unsigned char sec - BCD codification for sec, 00-59
    - o unsigned char min - BCD codification for min, 00-59
    - o unsigned char hour - BCD codification for hours, 00-24

Besides these data structures there is an additional data type that is retrieved during RTCC initialization that captures the state of the RTCC.

- **RtccRes** - result obtained from RTCC initialization
    - o RTCC_CLK_ON - success, clock is running
    - o RTCC_SOSC_NRDY - SOSC not running
    - o RTCC_CLK_NRDY - RTCC clock not running
    - o RTCC_WR_DSBL - WR is disabled

- **RtccInit (void)**-This function initializes the RTCC device. It starts the RTCC clock, enables the RTCC and disables RTCC write, disables the Alarm and the OE, clears the alarm interrupt flag and disables the alarm interrupt. The function returns the RtccRes value described above.

- **RtccGetClkStat ()** - This retrieves the status of the RTCC via RtccRes value. It does not perform any initialization.

- **RtccSetTimeDate (RtccTime, RtccDate)** –sets the RTCC internal registers to designated date, time variable structures.

- **RtccGetTimeDate (& RtccTime, & RtccDate)** – retrieves the date/time from RTCC internal registers to designated date, time variable structures.

- **mOSCEnableSOSC ()** – this is a critical i configuration setting to insure the device the secondary oscillator is enabled, called once at the start of Main.

This function is not part of the Microchip library but was written as part of the book's provided API to facilitate viewing of RTCC date, time variable structures in ASCII. We will use it to interface with the LCD.

- **RTCCProcessEvents ()** - The function grabs the current time from the RTCC and translate it into strings. Must be called periodically to refresh time and date strings _time_str and _date_str. Each string is 16 ASCII characters

Experiment 1 -Date and Time Setting and Operation

Let's get started using these concepts with our first experiment. In this experiment we will construct our external 32 KHz oscillator and connect it to the SOSCI and SOSCO pins of the Microstick II. It addition we will add a RED LED to indicate oscillator failure or RTCC failure and then a final green led that will blink every ½ second indicating successful RTCC operation. Open Chapter 12 and navigate to date time folder, exercise 1 with MPLAB X. Let's examine the code. Most of the action is in Main.c.

Main starts out with reference libraries and variable initializations.

```
#include "system.h"
#include "delay.h"
char _time_str[16] = "           ";     //
char _date_str[16] = "           ";     //
rtccDate dt;
rtccTime tm;
rtccRes res, clkStat;
void RTCCProcessEvents(void);
```

Figure 12-4: main code variable and reference declaration

In Figure 12-4 references to System.H and Delay.h are made. The RTCC data and time structures are declared, as well as RtccRes. The output string for RTCCProcessEvents (), (_time_str [16], and _date_str [16]), are declared and the functional prototype for RTCCProcessEvents () is also declared.
In the next code snippet we will see that the initialization and test moving through several discrete steps.

1. Enable secondary oscillator mOSCEnableSOSC ()
2. Initialize RTCC using RtccInit () and test res result to see is secondary Oscillator is stable and operational.
3. Check that RTCC is running before setting time and data using clkStat=RtccGetClkStat()
4. Write required data and time and set RTCC with these values using RtccSetTimeDate(tm.l, dt.l);

Note for steps 2 and 3 the red LED is lit if an error occurs.

```
mOSCEnableSOSC();
PORTSetPinsDigitalOut(IOPORT_B, BIT_3 | BIT_15);
mPORTBClearBits(BIT_15 );
Delayms(200);
//       It usually takes 4x256 clock
//cycles (approx 31.2 ms) for the oscillator signal to be available to the RTCC
        res =RtccInit();
   while (res==RTCC_SOSC_NRDY)
           {
        mPORTBSetBits(BIT_15 );     //turn on Led  and halt if no SOSC
           res =RtccInit();
}
   mPORTBClearBits(BIT_15 ); //clear led
 clkStat=RtccGetClkStat();
while (clkStat !=RTCC_CLK_ON)
{
    mPORTBToggleBits( BIT_15 );//turn on Led  and wait until Clock RTCC ready
    Delayms(500);
    clkStat=RtccGetClkStat();
}
  mPORTBClearBits(BIT_15 ); //clear led
        // let's start setting the current date
        {       tm.l=0;
                tm.sec=0x30;
                tm.min=0x07;
                tm.hour=0x10;
                dt.wday=2;
                dt.mday=0x16;
                dt.mon=0x01;
                dt.year=0x07;
                RtccSetTimeDate(tm.l, dt.l);
        }
```

Figure 12-5: RTCC init and set up.

The final stage the output RTCC is set for ½ second pulse rate and the clock is read and results are processed into the data time strings. Place a breakpoint in code as indicated and watch for strings to examine visually time data settings and changes.

```
51      RtccSelectPulseOutput(1);        // select the seconds clock puls
52      RtccOutputEnable(1);             // enable the Output pin of the
53      while(1) {
54
55      RtccGetTimeDate(&tm, &dt);
56          RTCCProcessEvents();
57        Delayms(2000);   //set brekpoint and view tm and dt
58
59      }
```

Watches

Name
 _time_str
 _date_str

Figure 12-6: Main loop with watch setting

With a successful operation we can breakpoint and examine the _time_str and _date_str contently and see the RTCC results directly.

The external oscillator consists of two 33 pF ceramic capacitors and a 32.768 KHz tuning crystal.

(2) 33pF ceramic capacitors Mouser 80-C315C330J1G5CA

(1) Clock Crystal CY32.76 Jameco part #14584

Equivalent parts are readily available from a number of vendors. Please wire the crystal and capacitor as closely as possible to the Microstick II pins and each other.

Figure 12-7: RTCC Prototype

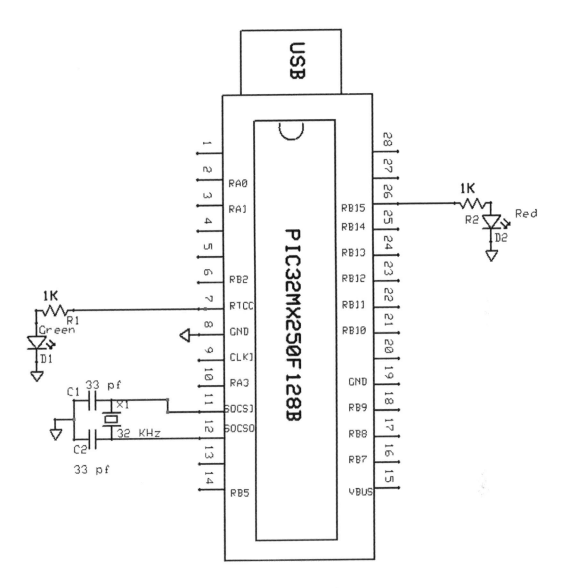

Figure 12-8: Experiment 1 schematic

The prototype is shown in Figure 12-7. Note the author constructed the external oscillator components by soldering to a small perforated board. A close up of that construction is shown in Figures 12-9 and 12-10. The board uses a 2 pin .100 male header mounted to bottom of the assembly to connect the assembly to the SOSCI and SOSCO pins of the PIC32MX. A single ground wire was used to connect with prototype ground. Other construction schemes are possible.

Figure 12-9 Top view of external crystal and dual cap construction

Figure 12-10 Bottom View of construction shows one ground wire and 2 pin header for crystal

Navigate to Chapter 12 date time folder, exercise 1 project and open with MPLAB X. Select project, make sure Microstick II is connected and is populated with hardware as shown in Figure 12-8. Run debug project. If red LED is lit there is a problem with the external oscillator wiring. Please fix before restarting. If a green LED is blinking –you are in business! The RTCC is operational. Set a breakpoint at the Delaymsec function in the loop and then check the running RTCC date time using watch variable _date_str, _time_str as shown

Experiment 2 Date and Time with LCD

The above functions work directly with the RTCC; however using a LCD with the RTCC lets you view current clock date time settings on the fly. We will be reusing the LCD from Chapter 5 and its associated libraries to work with the RTCC as shown in Experiment 1. Instead of needing breakpoints we will have the LCD display the contents of _date_str and _time_str in lines 1 and 2 of our LCD in real time.

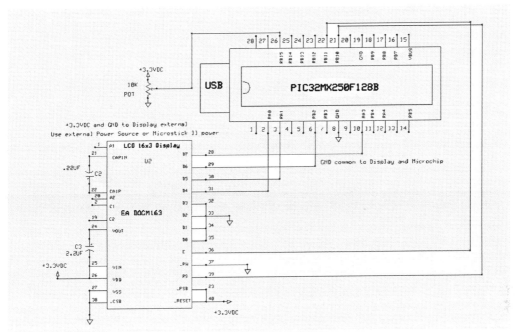

Figure 12-11: RTCC with LCD

The software changes to exercise 1 are minimal. We will just reuse our LCD and delay functions introduced in Chapter 5.

1. Add Lcd.h reference to main
   ```
   #include "system.h"
   #include "lcd.h"
   #include "delay.h"
   ```
2. Initialize and then clear display
   ```
   LCD_Initialize();
   clear_display();
   ```
3. Output two strings data and time to LCD display.

```
while(1) {

    RtccGetTimeDate(&tm, &dt);
    RTCCProcessEvents();
    position_cursor(0);              //position cursor to row 1
    write_array_LCD( (char*) _date_str,15);
    position_cursor(16);             //position cursor to row 2
    write_array_LCD( (char*) _time_str,15);
    Delayms(1000);  //

}
```

Navigate to Chapter 12 date time LCD folder, exercise 1 with MPLAB X. Select project, make sure Microstick II is connected and is populated with hardware as shown in Figure 12-11.

Run debug project. If red LED is lit there is a problem with the external oscillator wiring. Please fix before restarting. If a green LED is blinking –your RTCC is functional. If the LCD is displaying Date and time you are in business! Otherwise check your LCD wiring.

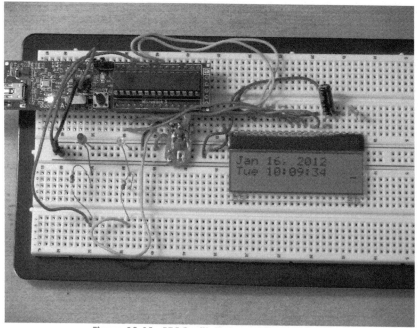

Figure 12-12: RTCC with Date and Time on LCD

Experiment 3 Alarm and Interrupt

The RTCC supports an alarm feature that allows the RTCC to interrupt the PIC32MX when an alarm condition occurs. By definition this is when the RTCC time and date match the alarms time and date. In this exercise we will use the existing prototype setup from the previous experiment without adding any additional hardware. The code will make generous use of the Microchip Peripheral Library to set up alarms and interrupt. The alarm condition will be handled in the prototype by adding an alarm notification on the 3rd line of the LCD when an alarm occurs.

Once an RTCC alarm occurs it can be configured to repeat with any of settings:
- every second
- every 10 seconds
- every minute
- every 10 minutes
- every hour
- every day
- every week
- every month
- every year

The other setting for alarm is repeat count that is user configured where the existing alarm condition will repeat as often as the user required. In our case we will set the alarm to occur every second, no user repeats, and upon first occurrence we will turn off the interrupt, in fact, only allowing one alarm to occur.

The RTCC interrupt service performs the function required for alarm notification. Code snippets for set alarm and RTCC interrupt and main code interaction are shown in Figures 12-14 and 12-15. In our ISR case we simply set a flag (alarm flag) to alert the main code, that an alarm has occurred, and to have main code output the message "Alarm occurred! " on the bottom row of the LCD. Also in response to the alarm main code will clear the alarm flag and disable RTCC interrupts so only one interrupt occurred. The prototype should look like Figure 12-13 once an alarm has occurred, the alarm message will be issued. Normal time and data update to the LCD will continue as before without interruption.

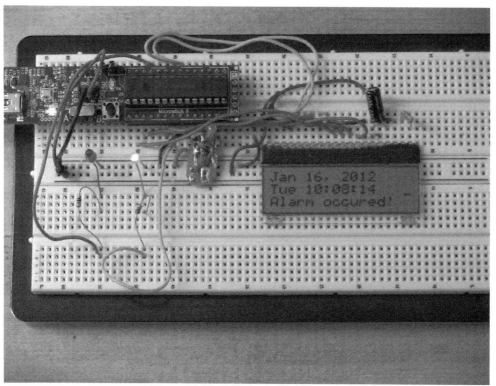

Figure 12-13: Prototype Alarm Operation

The interrupt service routine is as follows; we clear the RTCC interrupt flag and set a software flag alarm flag to signal the main routine.

```
void __ISR(_RTCC_VECTOR, ipl4) RtccIsr(void)
{
        // once we get in the RTCC ISR we have to clear the RTCC int flag
        INTClearFlag(INT_RTCC);
        alarmflag =1;

}
```

Figure 12-14: RTCC ISR

The main loop within the code does the normal job of updating date and time to rows one and 2 of the display. Also as part of the loop it checks alarm flag. If set it then displays the alarm message on row three of the LCD and then turns off the RTCC interrupt, so no further interrupts can occur.

```
        while(1) {

            RtccGetTimeDate(&tm, &dt);
            RTCCProcessEvents();
            position_cursor(0);              //position cursor to row 1
            write_array_LCD( (char*) _date_str,15);
            position_cursor(16);             //position cursor to row 2
            write_array_LCD( (char*) _time_str,15);

            if (alarmflag ==1) {
                position_cursor(32);         //position cursor to row 3
        write_string_LCD((char*)Alarm1);     //outputs Alarm Message 3 to display
    // ok, we're done
                INTEnable(INT_RTCC, INT_DISABLED);// disable further RTCC interrupts
                alarmflag =0;
            }
            Delayms(1000);   //
        }

    }
```

Figure 12-15: Main loop

Let's review how the alarm is set up in the code and how interrupts are enabled. First of all we add to the date, time structure to allow setting the RTCC alarm.

```
rtccDate dt, dAlrm;
rtccTime tm, tAlrm;
```

We also need to enable multi-vector interrupts using the library function

```
INTEnableSystemMultiVectoredInt();
```

The next step is to initialize the alarm (see Figure 12-16). To do this we read the existing RTCC date and time and then set the Alarm date and time with this same value (but offsetting seconds offset by 2 for a downstream alarm). The first test prior is a do loop where we actively read RTCC data and time make sure (as a precaution) that an internal RTCC rollover has not occurred to prevent contaminating the alarm setting. The Chime function for alarm is enabled but not used by our hardware. The user repeat count is zero, and the alarm repeat is set for every second.

The library function **RtccSetAlarmTimeDate (tAlrm.l, tAlrm.l)** sets the alarm value and the **RtccAlarmEnable ()** enables the alarm.

```
do
{
        RtccGetTimeDate(&tm, &dt);   // get current time and date
}while((tm.sec&0xf)>0x7);           // don't want to have minute or BCD rollover

tAlrm.l=tm.l;
dAlrm.l=dt.l;

tAlrm.sec+=2;                                  // alarm due in 1 secs

RtccChimeEnable();                             // rollover
RtccSetAlarmRptCount(0);                // we'll get more than one alarm
RtccSetAlarmRpt(RTCC_RPT_SEC);          // one alarm every second
RtccSetAlarmTimeDate(tAlrm.l, dAlrm.l); // set the alarm time
RtccAlarmEnable();                      // enable the alarm
```

Figure 12-16: Setting up alarm

Finally once the alarm has been configured, we double check that alarm is set and then configure the RTCC Interrupt.

```
        while(!RtccGetAlarmEnable());// check and wait thata alarm is enabled
// enabling/disabling the RTCC alarm interrupts
// set the RTCC priority in the INT controller
INTSetVectorPriority(INT_RTCC_VECTOR, INT_PRIORITY_LEVEL_4);
// enable the RTCC event interrupts in the INT controller
INTEnable(INT_RTCC, INT_ENABLED);
```

Figure 12-17 Configuring the RTCC Interrupt

Navigate to Chapter 12 folder data time alarm with LCD using MPLAB X. Connect the Microstick II and run debug project. You should see the results indicated by Figure 12-13 in a couple of seconds.

Review of RTCC:

- The RTCC provides the PIC32MX with a real time clock and calendar capability that support a hundred year Calendar.

- RTCC peripheral operation only requires a software setup and then runs independent of PIC32MX in hardware.

- RTCC support alarms through interrupt.

- There are a number of RTCC SFR settings to initialize clock. These setting are done using BCD.

- An external 32 KHz Crystal and two 33 pF capacitors are required to support RTCC.

Review of source code:
- Code reused the delay and LCD libraries.

- Significant use of Microchip Peripheral Libraries for RTCC

- Alarms are serviced through the RTTC interrupt service routine.

Exercise

1. Explain how the RTCC performs its counting functions based upon the 32Khz oscillator

2. What function is used to enable the secondary oscillator?

3. Was PPS used in any of the RTCC setups?

4. Reprogram data and time with different setting and monitor results

5. Reprogram the alarm with different setting and monitor results.

6. Discuss how you might use the alarm repeat function.

7. Discuss how you might use the chime function.

8. What other types of application functions can be implemented to support different alarm actions within the interrupt service routine?

Chapter 13 –Executing Arduino Code on PIC32

The Arduino is the de facto standard tool for developing microcomputer applications within both the hobbyist and educational communities. It provides an open-source hardware (OSH) environment based on a simple microcontroller board, as well as, an Open Source (OS) development environment for writing software for the board.

In this chapter we will introduce an approach that allows Arduino code to be configured for execution with the PIC32MX250F128B small outline 32 bit microcontroller, using the MPLAB X environment and the Microstick II programmer/debugger. We will work through four Open Source examples. Each example will add additional capability to our PIC32MX Arduino library, with the final example representing the complete baseline.

Your own reasons in working through these exercises, and using this approach, will depend on your own personal needs and background. Perhaps, as a long term Arduino user, you may want to explore a new processor performance option with your existing Arduino code base, or perhaps, you may want to take advantage of and/or gain experience with the Microchip advanced IDE development tools and debug with your existing Arduino code. Using the approach and the library covered in this Chapter all of these goals are easily achieved.

The Arduino source code will require some minor level of modification at the main function level to be compatible with Microchip XC32 'C' environment. Once this modification is done and "wrapped" into a MPLAB X project template (that includes a core Arduino library) the PIC32MX can then operate using Arduino code. This approach has worked successfully for several basic Open Source Arduino code examples. It has limitations, in that the implemented software provides only a subset of basic Arduino library functions; however this could change in the future. Another limitation is that target code must be "wrapped" for execution in a 'C' language environment and that any additional library imports beyond basic capability must be configured as 'C' files; a final limitation is that some of the existing Arduino library functions will need to be recast to run under a pure 'C' environment. You will experience these limitations firsthand in the ongoing code examples and be able to judge for yourself the applicability of this approach and its practicality for your future Arduino developments.

An assumption is made that the reader is familiar with Arduino OSH and OS software environments. If not please refer to the extensive material on the web associated with the Arduino.

http://arduino.cc/en/Reference/HomePage

Using this approach will allow you to contrast and compare Arduino OSH and OS software environment to the Microchip PIC32MX250F128B hardware environment, XC32 'C' compiler, and free Microchip MPLAB X in a direct way. As mentioned earlier, only a small core software subset of the larger Arduino Software Library functionality is represented. This implementation is captured in the project examples as Arduino.h, Arduino.c and other 'C' library reference files. The subset was purposely constrained to allow the four basic Open Source code examples from the Arduino Tutorial page to execute under the Microchip PIC32MX environment. See following web site for the examples.

http://arduino.cc/en/Tutorial/HomePage?from=Main.LearnArduino

1. **Blink**: Turn an LED on and off.
2. **DigitalReadSerial:** Read a switch; print the state out to a Serial Monitor.
3. **AnalogReadSerial:** Read a potentiometer; print its state out to a Serial Monitor.
4. **Fade:** Demonstrates the use of analog PWM output to fade an LED.

These examples illustrate what can be done to execute modified Arduino code in the Microchip environment. The reader is encouraged to expand on this core subset, as needed, to address even larger and more extensive Arduino code needs. We will introduce an approach to do this in the second exercise.

Core Subset of Arduino Library Functions
The Arduino fortunately emulates basic 'C' code in majority of its syntax and operations making adaption to Microchip XC32 'C' compiler fairly straightforward. The selected subsets of implemented Arduino library functions are:

- Digital I/O
    - pinMode(pin, mode)
    - digitalWrite(pin, value)
    - int digitalRead(pin)
- Analog I/O
    - int analogRead(pin)

- analogWrite(pin, value) - *PWM*
- **Time**
  - delay(ms)
  - delayMicroseconds(us)
- **Serial Communication**
  - Serial.begin(speed)
  - Serial.println(data)

- **Constants**
  Constants are particular values with specific meanings.
  HIGH | LOW
  INPUT | OUTPUT

- **Data Types**
  Variables can have various types, which are described below. These variable types are automatically recast with the XC32 compiler for basic PIC 32 bit machine.

   char
   byte
   int
   unsigned int
   long
   unsigned long
   float
   double
   string
   array

The PIC32MX250F128B Hardware Model

The Arduino OSH and Software library constraints its processor pin assignments for digital, analog, serial or PWM operations specific to the Arduino microcontroller device. The PIC32MX250F128B has its own unique microcontroller pin profile for digital, analog, serial and PWM. We map the Arduino designated D0 to D13 and A0 to A5 processor pins to the corresponding PIC32MX's unique pin profile to insure commonality of operation for the library functionality. The PIC32MX allows for its multi-functional pins to be reconfigured for different functions based upon PPS (Programmable Peripheral System—see chapter 9). A table is provided. Review this table for pins identification for prototyping. Note that D7 is only supported with MX1 parts. MX2 parts like our PIC32MX250F128B have a pin limitation due to a built-in USB capability that uses D7 (pin 15) for VBUS. However, there is a configuration setting in our arduino.h library file to accommodate both part types: #define MX1. It is set to 0 for MX2 parts (default). Set this to one if you are using an MX1 part.

As a final note keep in mind that the PIC32MX1 and PIC32MX2 parts are +3.3V DC parts versus the Arduino which is +5V DC part. Some pins on each of the MX1 and MX2 series are however +5V tolerant, you need to consult the data sheet available from the following web page before making a +5 VDC connection to a MX1/MX2 part.

http://www.microchip.com/wwwproducts/Devices.aspx?dDocName=en557425

Connecting +5 VDC to a MX1/MX2 pin that is non-tolerant of +5VDC can destroy the pin's functionality.

Designator (pin)	ARDUINO	PIC32 pin	PIC32MX250F128B
D0 (0)	RXD/PD0	22	RBP11\UART2RX
D1 (1)	TXD/PD1	21	RBP10\UART2TX
D2 (2)	INT0/PD2	16	INT0\RPB7
D3 (3)	INT1/PD3	24	INT1\RPB13
D4 (4)	T0/PD4	12	T1CLK\RPA4
D5 (5)	T1/PD5	11	T2CLK\RPB4
D6 (6)	AIN0/PD6	9	RPA2
D7 (7)	AIN1/PD7	15	RPB6 -available only on MX1 parts
A0 (14)	ADC0/PC0	2	AN0\RA0
A1 (15)	ADC1/PC1	3	AN1\RA1
A2 (16)	ADC2/PC2	4	AN2\RB0\PGED1
A3 (17)	ADC3/PC3	5	AN3\RB1\PGEC1
A4 (18)	ADC4/SDA/PC4	6	AN4\SDA2\RB2
A5 (19)	ADC5/SCL/PC5	7	AN5\SCL2\RRB3
D8 (8)	ICP/PB0	10	IC2\RPA3
D9 (9)	OC1A/PB1	26	OC1\RB15
D10 (10)	OC1B/SS/PB2	18	OC3\RB9
D11 (11)	MOSI/OC2/PB3	17	SDO1\OC2\RPB8
D12 (12)	MISO/PB4	14	SDI1\RPB5
D13 (13)	SCK/PB5	25	SCK1\RB14

Figure 13-1: PIC32 Library supported pin out

Arduino Blink Example #1

The Arduino code example is as follows: Wire an LED to through a 1K resistor to pin 13 (D7) of Arduino. An output pin is configured to drive an LED using pinMode () function under setup (), and then under loop () this output is set high and then low, using digitalWrite () and delay () functions, to blink the LED. The Community Open Source Arduino code is as follows:

```
/*
Blink
Turns on an LED on for one second, then off for one second, repeatedly.
This example code is in the public domain.
*/
// Pin 13 has an LED connected on most Arduino boards.
// give it a name:
int led = 13;
// the setup routine runs once when you press reset:
void setup() {
// initialize the digital pin as an output.
pinMode(led, OUTPUT);
}
// the loop routine runs over and over again forever:
void loop() {
digitalWrite(led, HIGH); // turn the LED on (HIGH is the voltage level)
delay(1000); // wait for a second
digitalWrite(led, LOW); // turn the LED off by making the voltage LOW
delay(1000); // wait for a second
}
```

Figure 13-2: PIC32 Example 1 Code Modifications

The Open Source example uses D13 or physical pin 13 on the Arduino. In relation to the PIC32MX the D13 is physical pin 25. Pin 25 will be used in prototyping wiring.

Now, let review and understand the PIC32 project template and its associated "wrapping functions". The Arduino uses two principal functions setup () to initialize the system and loop () to run a continuous execution loop. There is no Main function. Using the Microchip XC32 'C' compiler we are constrained to having a Main function. The Arduino setup () and loop () functions can be accommodated, but only as part of an overall template Main "wrapping" function. So within our PIC32 template we accommodate this as follows:

```
//***********************************************
main() {

// the setup routine runs once when you press reset:
setup();
// the loop routine runs over and over again forever:
while(1){
loop();
   };
//end of code              //
}
```
Figure 13-3: Main Wrapper Code

This piece of code is a small but essential part of the template. Note that in this critical wrapping function setup () is called once as in Arduino and loop () is configured to be called continuously (simulating the loop () function in Arduino) through the use of a while loop in Main.

The second critical wrapping function for our template is the use of 'C' header files at the beginning of the code. The XC32 'C' compiler uses the C compiler directive "#include" reference files within the Main code. Arduino uses "import", which is a similar construct that is used in higher level languages like Java and Python, which cannot be used by XC32 'C'.

The two include files necessary for our first example is as follows:

```
#include "system.h"
#include "Ardunio.h"
```
Figure 13-4: Library Wrapper Code

System.h references all the critical Microchip library functions supporting the PIC32MX250F128B. The Ardunio.h provides the Arduino specific library function set. Given these two key "wrapper" aspects, where does the Arduino code go? This is best illustrated with a side by side comparison between Arduino Code and its Microchip equivalent. The Arduino code is essentially positioned between the wrapper codes as part of the Main function.

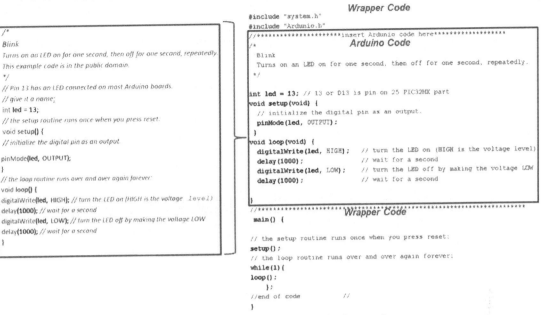

Figure 13-5: Example 1 "Blink" Side by Side Code Comparisons

This approach allows for Arduino code to execute on a PIC32 within a MPLAB X environment. Note that the Arduino code void setup () now appears as void setup (void), and void loop () appears as void loop (void). This is a minor inconvenience but again necessary for our 'C' environment syntax for 'C' prototype function definitions. Once the code is successfully compiled the environment allows you to have access to the entire built-in tool suite of the MPLAB X and its debugger tool suite.

Running Example 1 Code

Configure the prototype as per the following schematic, both schematic and prototype are shown.

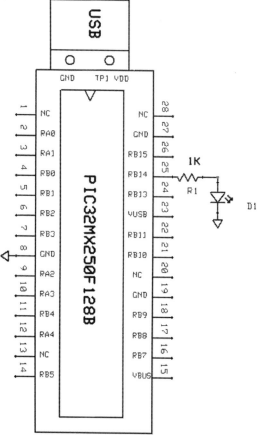

Figure 2: Exercise 1 Schematic

Figure 3 Exercise 1 Prototype

Navigate to the Chapter 13 folder project Arduino Blink using MPLAB X, and open project Blink .Make sure Microstick II prototype is configured with LED at pin 16. The MPLAB X project IDE should appear as follows:

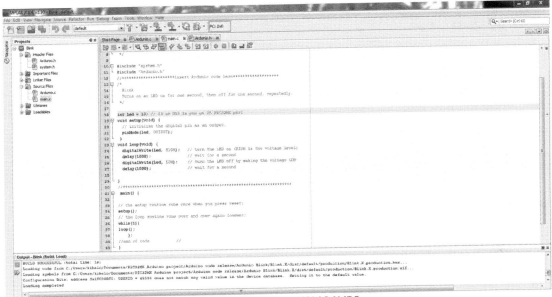

Figure 4-8: Exercise 1 Blink MPLAB X IDE

Start the Project Debug under MPLAB X. You should see the LED blink as per the Arduino source code, if not recheck the wiring, you must have a mistake.

Arduino Digital Serial Read Example #2

Wire a button to pin 2 (D2) of the Arduino hardware.

In this program the setup function initializes serial communications, at 9600 bits of data per second, between your Arduino and your computer with the line:
Serial.begin (9600);

Next, we initialize digital pin 2, the pin that will read the output from your button, as an input:

- pinMode (2,INPUT);

In loop if the button is not pressed, +5 volts through a 10K resistor will be read as high, and when it is pressed, the input pin will be connected to ground, it will be read as low. This is a digital input, with high representing no switch depression and low indicating switch depression.

During loop a variable to hold is created for the information coming in from the switch. This variable is sensorValue, and set it to equal whatever is being read on digital pin 2.

- Int sensorValue = digitalRead (2);

Once the Arduino has read the input, it will print this information back to the computer as a decimal value with the command Serial.println (). The Open Arduino code is as follows:

```
/*
DigitalReadSerial
Reads a digital input on pin 2, prints the result to the serial monitor
this example code is in the public domain.
*/
// digital pin 2 has a pushbutton attached to it. Give it a name:
int pushButton = 2;
// the setup routine runs once when you press reset:
void setup() {
// initialize serial communication at 9600 bits per second:
Serial.begin(9600);
// make the pushbutton's pin an input:
pinMode(pushButton, INPUT);
}
// the loop routine runs over and over again forever:
void loop() {
// read the input pin:
int buttonState = digitalRead(pushButton);
// print out the state of the button:
Serial.println(buttonState);
delay(1); // delay in between reads for stability
}
```

Figure 13-9: PIC32 Example 2 Code Modifications

We will use pin D2, pin 16 for PIC32MX, digital input (just like Arduino Open Source). The Microstick II does not come equipped with its own complete serial USB interface like the Arduino, so for a serial input we need to reconstruct Chapter 9 Exercise 1 USB Serial using Tera Term Software for terminal interface and the ACRONAME USB-to-UART S27 interface.

The schematic is shown in Figure 13-12. The Serial.begin () and Serial.println () Arduino code will be recast as Serial_begin () and Serial_println (). The '.' extension used with Serial in Arduino is typical of higher level language (i.e.

Java) extensions, where the Serial is a class and the println () and begin () are methods of that class.

In the 'C' environment using '.', there are no classes, and the syntax is typically used for structures. In it a minor nuance, but to circumvent this we replace the '.' notation with the '_' to clearly indicate to 'C' compiler that these are functions. You will need to modify your Arduino serial function calls to successfully compile under XC32. Other than this everything else is unchanged.

To accommodate Serial_begin (), and Serial_println () the project will incorporate two new files from our existing library into the template: CONU2.h and CONU2.c, recast with the new Arduino functions. We are also using 2400 baud versus 9600 baud, because that was what our Chapter 9 UART exercises used. Serial_begin () always initializes UART 2 of the PIC32 to 8N1 (8 bit data , no parity, and 1 stop bit) .The Serial_begin () can accommodate other baud as well within the constraint of the 8MHz CPU clock setting. Higher baud rate may require reconfiguring the CPU clock rate in System.h.

The Serial_println () is implemented to handle only integer arguments at this point in the design, as opposed to Serial.println () which handles both integer and float automatically.

Our approach uses this add on modular approach to facilitate Arduino library extension without necessarily "over sizing" the core library functional files Arduino.h and Arduino.c. In this matter we can add our existing C libraries with only slight modification to address the Arduino functional call specifics.
The top part of the wrapper code adds a new reference file while the bottom wrapper code remains the same.

```
#include "system.h"
#include "Ardunio.h"
#include "CONU2.h"
```

Figure 13-10: Example 2 Library wrapper code

Arduino Open Source Code

```
/*
DigitalReadSerial
Reads a digital input on pin 2, prints the result to the serial monitor
this example code is in the public domain.
*/
// digital pin 2 has a pushbutton attached to it. Give it a name:
int pushButton = 2;
  // the setup routine runs once when you press reset:
  void setup() {
  // initialize serial communication at 9600 bits per second:
  Serial.begin(9600);
  // make the pushbutton's pin an input:
  pinMode(pushButton, INPUT);
}
  // the loop routine runs over and over again forever:
  void loop() {
  // read the input pin:
  int buttonState = digitalRead(pushButton);
  // print out the state of the button:
  Serial.println(buttonState);
  delay(1); // delay in between reads for stability
}
```

PIC32 Open Source Template

```
#include "system.h"          Wrapper Code
#include "Arduino.h"
#include "CONU2.h"
//*********************insert Arduino code here*************************
//DigitalReadSerial          Arduino Code
//Reads a digital input on pin 2, prints the result to the serial monitor
//this example code is in the public domain.

// digital pin 2 has a pushbutton attached to it. Give it a name:
int pushButton = 2;
// the setup routine runs once when you press reset:
void setup(void) {
// initialize serial communication at 9600 bits per second:
Serial_begin(2400);
// make the pushbutton's pin an input:
pinMode(pushButton, INPUT);
}
// the loop routine runs over and over again forever:
void loop(void) {
// read the input pin:
int buttonState = digitalRead(pushButton);
// print out the state of the button:
Serial_println(buttonState);
delay(1); // delay in between reads for stability
}
//*********************************************************************
main()  {                    Wrapper Code
// the setup routine runs once when you press reset:
setup();
// the loop routine runs over and over again forever:
while(1){
loop();
  };
//end of code              //
}
```

Figure 13-11: Example 2 "DigitalReadSerial" side by side comparison

Running Exercise 2 Code

Wire up prototype as shown in Figure 13-12. Both schematic and prototype are shown.

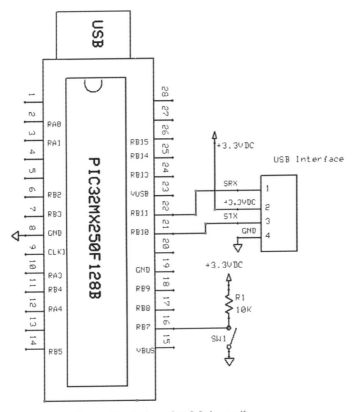

Figure 13-12: Exercise 2 Schematic

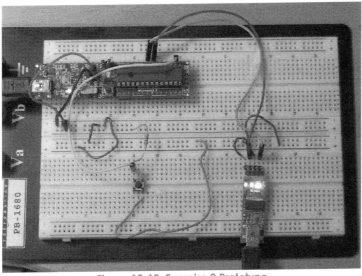

Figure 13-13: Exercise 2 Prototype

Navigate to the Chapter 13 folder project DigitalSerial using MPLAB X. Note the Modified Arduino Open Source in the Main.c with top and bottom wrapper code as illustrated earlier. You now can execute the original open source using full debug suite of MPLAB X.

Connect the Microstick II to PC USB. Use a second USB port and cable to connect to ARCONAME USB-to-UART interface. Next, open TERA TERM to an available com port and then set this port to 8N1 2400. In MPLAB X to start the debug process you must click on Project Debug icon. You should see the state of the button depression capture on the TERA TERM window as shown in the Figure 13-14, pushbutton state capture: button is 1, then depressed is 0 , and then released is 1. Press and release your button and verify window display results.

Figure 5: Exercise 2 Pushbutton state capture using Teraterm

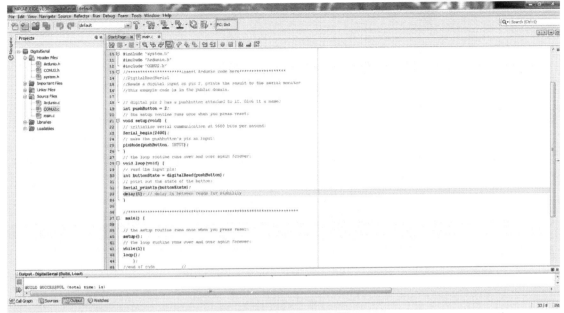

Figure 13-15: Example 2 MPLAB IDE

Arduino Analog Serial Read Example#3

Wire a potentiometer to A0 of the Arduino. The example illustrates how to read analog input using a potentiometer. By passing voltage through a potentiometer and into an analog input, it is possible to measure the amount of resistance produced by a potentiometer (or *pot* for short) as an analog value. In this example we will monitor the state of a potentiometer and report its value using a serial communication to a PC. A new function is introduced:

- int sensorValue = analogRead (Analog input);

The Open source Arduino code is as follows:

```
/*
AnalogReadSerial
Reads an analog input on pin 0, prints the result to the serial monitor.
Attach the center pin of a potentiometer to pin A0,

And the outside pins to +5V and ground.
This example code is in the public domain.
*/
// the setup routine runs once when you press reset:
void setup() {
// initialize serial communication at 9600 bits per second:
Serial.begin(9600);
```

}
// the loop routine runs over and over again forever:
void **loop**() {
// read the input on analog pin 0:
int **sensorValue** = analogRead(A0);
// print out the value you read:
Serial.println(sensorValue);
delay(1); // delay in between reads for stability
}

Figure 6: Example 3 "AnalogReadSerial" side by side comparison

This example is very similar to Example 2; the only change is reading and reporting of an analog value rather than digital to the serial monitor. The library function is analogRead (pin), where we will use analog designator A0. This is pin 2 of the PIC32MX.

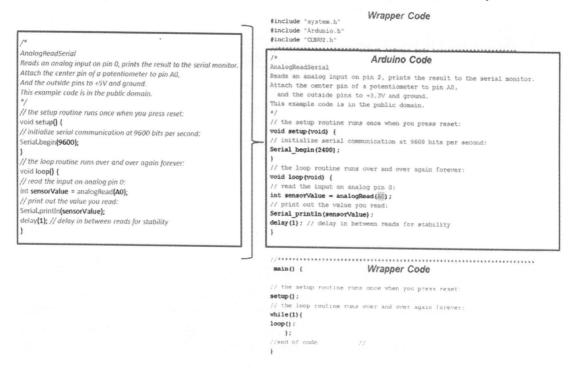

Figure 13-17: Example 3 "AnalogReadSerial" side by side comparison

Running Example 3 Code

Wire up prototype as shown in Figure 13-18, both schematic and prototype are shown.

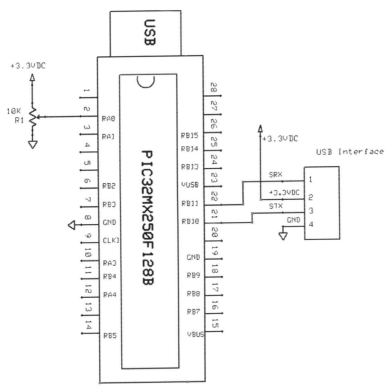

Figure 13-18: Exercise 3 Schematic

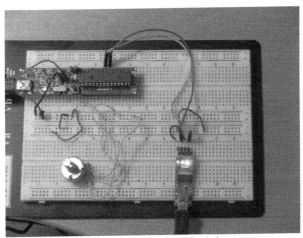

Figure 13-19: Exercise 3 Prototype

Beginner's Guide to Programming the PIC32

Navigate to the Chapter 13 folder project Analog Serial using MPLAB X. Note the Modified Arduino Open Source in the Main.c with top and bottom wrapper code as illustrated earlier. You now can execute the original open source using full debug suite of MPLAB X. Connect the Microstick II to PC USB. Use a second USB port and cable to connect to ARCONAME USB-to-UART interface. Open TERA TERM to available com port and set to 8N1 2400, in MPLAB X, click on Project Debug icon. You should see the state of the analog pot value capture on the TERA TERM window (see Figure 13-20). Change the pot settings and verify readings.

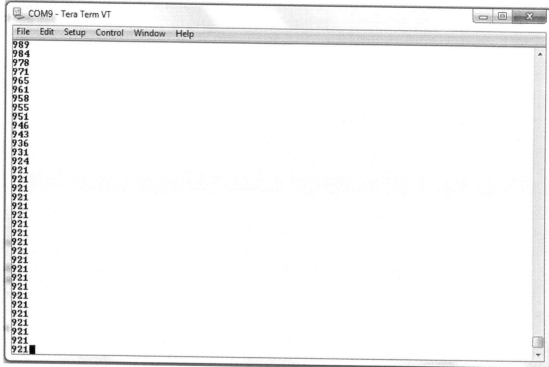

Figure 13-20: Exercise 3 Pot analog readings

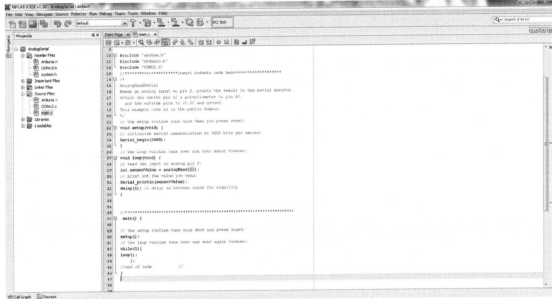

Figure 13-21: Exercise 3 MPLAB

Arduino Fading Example #4

This example illustrates how to control LED brightness using PWM. Specifically this demonstrates the use of the analogWrite () function in fading an LED brightness. AnalogWrite uses pulse width modulation (PWM), by turning a digital pin on and off very quickly, to create a fading effect. The Open Arduino code is as follows:

```
/*
Fade
This example shows how to fade an LED on pin 9
using the analogWrite () function.
This example code is in the public domain.
*/
int led = 9; // the pin that the LED is attached to
int brightness = 0; // how bright the LED is
int fadeAmount = 5; // how many points to fade the LED by
// the setup routine runs once when you press reset:
void setup() {
// declare pin 9 to be an output:
pinMode(led, OUTPUT);
}
// the loop routine runs over and over again forever:
void loop() {
// set the brightness of pin 9:
analogWrite(led, brightness);
```

```
// change the brightness for next time through the loop:
brightness = brightness + fadeAmount;
// reverse the direction of the fading at the ends of the fade:
if (brightness == 0 || brightness == 255) {
fadeAmount = -fadeAmount ;
}
// wait for 30 milliseconds to see the dimming effect
delay(30);
}
```

Figure 13-22: PIC32 Example 4 Code Modifications

The Open Source example uses pin D9 on the Arduino. D9 on the PIC32MX is physical pin 26 for PWM on the PIC32. D9 will be configured under PPS for PIC32MX Output Compare Perpherial1 and Timer 2. Chapter 11 in the book discusses PPS and the OC1 peripheral.

We will adhere to our on modular approach in facilitating the Arduino library extension for PWM. The top part of the wrapper code adds again a new reference file (PWM.h) to provide the analogWrite () the required PWM function while the bottom wrapper code remains the same

```
#include "system.h"
#include "Ardunio.h"
#include "CONU2.h"
#include "PWM.h"
```

Figure 13-23: Exercise 6 Library Wrapper Code

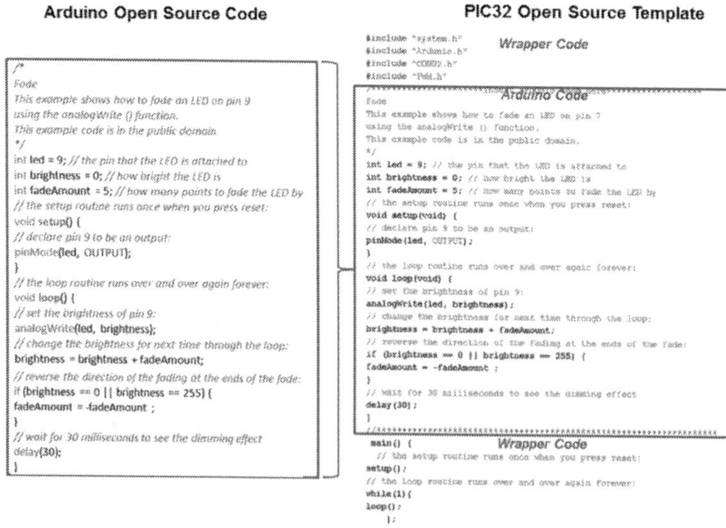

Figure 13-24: Example 4 "analogWrite ()" side by side comparison

Beginner's Guide to Programming the PIC32

Running Example 4 code

Wire up prototype as shown in Figure 13-25, both schematic and prototype are shown.

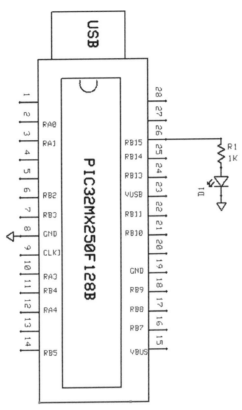

Figure 13-25: Exercise 4 Schematic

Figure 13-26: Exercise 4 Prototype

Beginner's Guide to Programming the PIC32

Navigate to the Chapter 13 folder project Analog Write using MPLAB X. Note the Modified Arduino Open Source in the Main.c with top and bottom wrapper code as illustrated earlier. You now can execute the original open source using full debug suite of MPLAB X. Connect the Microstick II to PC USB. Select Project Debug and you should see the LED proceed through fading cycle.

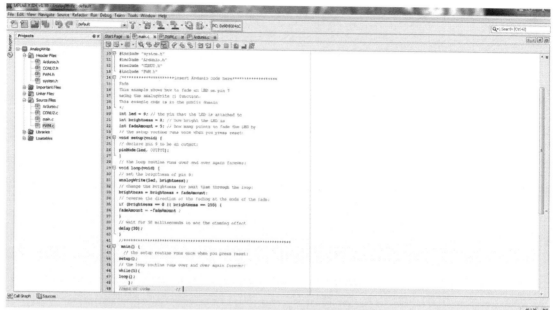

Figure 13-27: Exercise 4 MPLAB X view

Review of PIC32MX Key Arduino Porting Approach and Library Features:

An approach was covered in this Chapter to allow Open Source Arduino code to be executed by the PIC32MX250F128B using Microchip XC32 'C' Compiler, the MPLAB X environment and the Microstick II programmer/Debugger. The approach requires some minor level of modification to the Open source code to be compatible with Microchip XC32 'C' environment and use of "wrapper code" within the main function. The approach has been successfully demonstrated with several basic Open Source Arduino code examples. A limited capability for the Arduino reference library was developed. This capability is tailored to work with the PIC32MX250F128B microcontroller. The approach is extensible to other PIC32 bit devices, and also provides the necessary modularity to conveniently grow and enhance the library. The required reference files, within the top part of the "wrapper code" needed to obtain the total current library functionality are:

- **#include system.h** – provides the Microchip libraries, configuration

settings, and chip description for the PIC32MX250F128B
- **#include arduino.h** – provides the basic Arduino digital I/O, time, and analog read functions
- **#include CONU2.h** – provides the Serial functions using UART2
- **#include PWM.h** – provides the PWM function for analogWrite ()

The selected subset library functions are:

- **Digital I/O**
    - pinMode (int pin, int mode)- sets the designated pin to be either in INPUT or OUTPUT mode. Available D0-D13 pin designators are 0, 1, 2, 3, 4, 5, 6, 7 (MX1 parts only), 8, 9, 10, 11, 12, and 13. You can convert A0 to A5 to digital by using pin designators 14, 15, 16, 17, 18, and 19
    - digitalWrite (int pin, int value)- write value high or low to designated pin. Available pins are Available D0-D13 pin designators are 0, 1, 2, 3, 4, 5, 6, 7(MX1 parts only), 8, 9, 10, 11, 12, and 13. You can convert A0 to A5 to digital by using pin designators 14, 15, 16, 17, 18, and 19
    - int digitalRead(pin)- reads digital value(0 or 1)of associated pin and returns it as integer value. Available D0-D13 pin designators are 0, 1, 2, 3, 4, 5, 6, 7(MX1 parts only), 8, 9, 10, 11, 12, and 13. You can convert A0 to A5 to digital by using pin designators 14, 15, 16, 17, 18, and 19
    -
- **Analog I/O**
    - int analogRead(int pin)- executes a 10 bit ADC conversion on selected channel. The pin is configured for analog. The function returns an integer value representing the ADC value from 0 to 1024. Only one analog channel is enabled at a time, available pins are A0, A1, A2, A3, A4 ,and A5
    - analogWrite (int pin, int value) – PWM, accepts pins D9 and D10 , (duty cycle) can range from 0 (off) to 255 (on all the time). Uses PPS to set pin 9 to OC1 using Timer 2 and pin 10 to OC3 using Timer 3. PWM frequency under 100 Hz.
- **Time**
    - delay(int ms)- uses Timer1 to delay integer value (32 bits of count) in milliseconds.
    - delayMicroseconds(int us) –uses Timer1 to delay integer value (32 bits of count) in microseconds.

- **Serial Communication**
    - Serial_begin (int baud)- configures UART2 with baud rate and fixes operation for 8 data bits, 1 stop bit and no parity. Using PPS feature to set D1 pin 21 TX and D0 pin 22 RX.
    - Serial_println (int data) – limited to integer data. Print ASCII equivalent of integer data along with line feed and carriage return.

Exercise:

4) Try extending Serial function library using existing CONU2 code for
    a. int Serial.read () – read and return a integer from serial port
    b. Serial.print(data) – output data to serial port without LF and CR
5) Change CPU clock setting in system.h using configuration fuses from 8 MHz to 36 MHz to achieve higher baud rates then 2400 for Serial Function
6) Add a Arduino reference library function **EEPROM** using NVM code in Chapter 10 of book
7) Add a contributed Arduino reference library function **LCD 4 Bit** using LCD code in Chapter 6 of book
8) Add a Arduino reference library reference function **shiftOut (dataPin, clockPin, bitOrder, value)** using SPI functions covered in Chapter 10 of book
9) Investigate using #include <math.h> from Microchip C libraries to add following Arduino reference libraries:
    a. **Math**
        i. min(x, y)
        ii. max(x, y)
        iii. abs(x)
        iv. constrain(x, a, b)
        v. map(value, fromLow, fromHigh, toLow, toHigh)
        vi. pow(base, exponent)
        vii. sqrt(x)
    b. **Trigonometry**
        i. sin(rad)
        ii. cos(rad)
        iii. tan(rad)
10) Investigate using #include <stdlib.h> from Microchip C libraries to add following Arduino reference libraries:
    a. **Random Numbers**
        i. randomSeed(seed)
        ii. long random(max)
        iii. long random(min, max)

11) Investigate using Microchip peripheral library for PIC32MX Core Timer function to add following Arduino reference library:
    a. **Time**
        i. unsigned long millis()
12) Refer to Microchip Library of Applications (http://www.microchip.com/MLA) for other potential new Arduino library contribution ideas.

For comments, constructive criticisms, any detected problems, and new libraries additions please contact the author at thomas.kibalo@kibacorp.com.
The author will, post comments, and maintain the current baseline and errata for the PIC32MX Arduino library as a separate page on his web site at www.kibacorp.com

INDEX

2

25LC256, 187, 194, 197

A

AD1CON1, 101, 111
AD2CON2, 101, 111
AD3CON3, 101, 111
ADC, 6, 9, 33, 97, 121, 140, 201, 255
ANSELA, 50, 194
ANSELB, 50, 80, 194
Arduino, 8, 10, 230

B

Build Project, 24, 30

C

Change Detection, 84, 93, 94, 96
configuration settings, 46, 52, 67, 80, 254

D

Dashboard, 20, 21, 22, 24, 27, 54, 55, 61, 64
debounce, 79, 80, 82, 83
Debug, 10, 21, 27, 28, 29, 30, 57, 62, 66, 67, 76, 83, 238, 243, 248, 253
Debug Run, 30
Debug Tool, 21, 57
Delay function, 49, 67

E

Editor pane, 23
EEPROM, 7, 186, 187, 188, 189, 191, 192, 193, 194, 195, 197, 256

F

File pane, 23

H

HyperTerminal, 7, 155, 158, 162, 164, 171, 184

I

Interrupts, 6, 84, 95, 113, 127, 128, 134

L

LCD Display, 6, 113, 114, 212
LED Display, 5, 39

M

Make and Build All, 30
Make and Program, 30
Microchip, 2
Microstick II, 5, 25, 43, 48, 51, 53, 61, 66, 95, 100, 120, 135, 138, 170, 195, 221, 243, 254
MPLAB, 2
MPLAB X, 5, 10, 35, 44, 64, 74, 80, 182, 193, 196, 200, 205, 210, 216, 221, 223, 227, 230, 243, 248

N

Navigator pane, 23

O

Optimizing PIC32 Performance, 6, 141

P

Peripheral Library, 5, 16, 41, 42, 52, 166, 194, 206, 223
PIC, 2
PIC32MX250F128B, 5, 20, 52, 88, 96, 103, 127, 152, 164, 184, 194, 204, 212, 230, 254
PORTA, 40, 66
PORTB, 40, 51, 65, 70, 77, 80, 83, 93, 129
PPS, 7, 34, 38, 155, 164, 184, , 228, 232, 250, 255
PPSOutput, 194, 200, 204
push button, 70, 78
PWM, 7, 9, 33, 164, 198, 210, 231, 232, 250, 254

R

Real-Time Clock, 212, 213

Reset, 30, 37, 133, 139, 196
RS232 Communications, 7, 155
RS-485, 159, 163, 184
RTCC, 7, 9, 33, 212

S

SFR, 40
simulator, 31, 54, 55, 56, 61, 134
Solderless breadboard, 25, 26, 27, 43, 73, 124, 172
SPI, 7, 9, 33, 42, 164, 186, 256
Stopwatch, 53, 56, 58, 62, 67
Switch, 5, 37, 46, 69, 70, 71

T

Task pane, 23
Tera Term, 7, 158, 161, 164, 179, 180, 184, 239
Timer Peripheral Pairs, 6, 136
Timer2, 6, 134, 135, 137, 138, 140, 199, 200, 201, 205, 206, 210
Timer3, 199, 210

Timers, 6, 127, 128, 139, 140, 199
Tone Generation, 7, 198, 206, 209
TRIS, 40, 41, 48, 49, 70

U

UART to USB Bridge, 161
UART2, 7, 155, 167, 254, 255

V

Voltmeter, 6, 121, 122, 124

W

Watch, 59, 60, 66, 77, 83, 110, 133

X

XC32, 11, 15, 16, 20, 30, 67, 85, 134, 147, 152, 166, 230, 254

Made in the USA
Lexington, KY
23 June 2017